TOTAL

AN ESSENTIAL GUIDE TO CONTROLLING ANY

CRITTER

SORT OF CRITTER, FROM THE MERELY PESKY

CONTROL

TO THE OUTRIGHT DANGEROUS

Edited by
DON SEDGWICK

With illustrations by Wallace Edwards

The Lyons Press
Guilford, Connecticut
An imprint of The Globe Pequot Press

The Lyons Press is an imprint of The Globe Pequot Press.

Portions of this material have appeared in slightly different form in the following titles by Bill Adler, Jr.: *Outwitting Critters, Outwitting Deer, Outwitting Mice, Outwitting Neighbors,* and *Outwitting Squirrels.*

Printed in the United States of America

10 9 8 7 6 5 4 3 2 1

Library of Congress Cataloging-in-Publication Data

Total critter control : an essential guide to controlling any sort of
critter, from the merely pesky to the outright dangerous / edited by Don
Sedgwick ; with illustrations by Wallace Edwards.
    p. cm.
  ISBN 1-58574-850-1 (pb : alk. paper)
  1. Urban pests--Biological control. 2. Insect pests--Biological
control. 3. Wildlife pests--Biological control. I. Sedgwick, Don.
  SB603.3.T68 2003
  628.9'6--dc21
                        2003001228

CONTENTS

Contents

INTRODUCTION

Critters are everywhere and—from a distance—many of them are cute and interesting to watch. Deer, raccoons, and squirrels, for example, are favorites of children and adults. But when you have a garden, or if you're worried about Lyme disease, that doe and her fawn are suddenly not quite as welcome in your yard. If you've ever heard the loud and pathetic noises of raccoons mating in a tree outside your window (or worse, in your attic), then you know that these nocturnal prowlers can become real nuisances. And even the lovely little squirrels we see in the local park can become headaches when they take over your cherished bird feeder or decide to nest in your garage.

From the lone cabin at the edge of a remote forest to the split-level house in a suburban subdivision, no home is safe from these animal "invasions." Suburbs continue to expand and animals have to go *somewhere*. In many cases *they* were actually there first, and *we* are really doing the invading, or we've eliminated their environment, so they've moved into ours. Property lines don't mean anything to these critters. As the suburbs continue to spread, a greater number of homeowners will face the challenges of living with—or near—nature. And nature likes to remind us that it's still the boss by introducing us to its denizens.

Perhaps "introducing" is too friendly a word. Nature likes to barge in every now and then to remind us that there is more to the planet than just people and our things. No matter how hard we try to keep critters out of our houses, gardens, bird feeders, even swimming pools and cars, they are always a step ahead of us. Critters haven't necessarily grown any smarter, but over the years we've moved our homes closer to places where nature rules, and where we are just visitors. In many cases, they've simply moved in.

Houses can become desirable habitats for animals, especially if they are situated on formerly undeveloped areas that still retain streams, trees, and other vegetation. Our homes can offer suitable protection for nocturnal animals (ironically out of sight of our dogs and cats), and we unwittingly offer them an excellent food supply. Raccoons, skunks, and opossums are especially interested in discovering the benefits of urban living. Garbage cans, barbecue remnants, and even pet food left outdoors are manna to a skunk or raccoon. Meanwhile, our vegetable gardens and orchards are favorite desserts for deer. It's hard to blame these visitors when it appears to them that we have invited them into our homes and yards.

How do homeowners cope with these problems? Sometimes animal invasions are just annoyances. Rabbits will eat a few of your prized strawberries; a colony of spiders will take over your garage workshop; or pigeons will sit cooing under the soffits of your roof. Annoyance alone may be a good enough reason to tackle the problem, particularly if it's chronic. We've known bird-lovers to get absolutely manic about trying to keep squirrels out of a backyard feeder. But we can usually overlook the occasional inconvenience of cleaning up an overturned garbage container in the morning.

Sometimes, however, animal invasions are more than just annoyances. They can be absolutely dangerous. Coyotes and alligators have a fondness for small dogs, for example. Roaches, ants, and mice are all potential disease carriers. Deer may carry tiny ticks that also harbor disease, while foxes are prone to rabies. Unfortunately, there is no sure way to tell by merely looking at an animal whether it has rabies, except that rabid animals often exhibit no fear of humans—and even then they may simply be hungry. As a precaution, adults and children should stay well away from any wild animals, particularly those that appear unafraid or sick.

Critter invasions also pose a danger to your pets. Pets should obviously be kept well away from wild animals, and they should always be vaccinated against rabies. Since cats frequently roam more freely than dogs, and they enjoy hunting other animals, they often face the greatest risk of being bitten by a rabid animal. In addition to contracting the disease itself, a cat can also pass along rabies to its owner. At this point, the challenge of dealing with critters has obviously moved from the annoyance stage to that of a serious problem.

Fortunately, the best techniques for controlling unwanted wildlife are also the ones that don't injure animals. Harming or killing an animal is not only inhumane, it is guaranteed to make you feel lousy. From a practical point of view, it may not even be helpful. Dealing with the real underlying problem is always more useful. Besides, where there is one gopher, there are bound to be many more. Assume there is a partner for any creature that's bothering you—and probably a family tucked away in your yard as well. Rather than pulling out the heavy artillery, here are some of the more practical critter control suggestions you'll find in the following pages.

1. *Avoid pesticides and poisons.* You may kill the creatures you want to get rid of (or at least some of them), but you will also kill other animals you don't

wish to harm. Animals that are resistant to the poisons—or able to outsmart them—will be back in greater quantity and strength. The infamous cockroach that could withstand a nuclear blast has other equally hardy critter friends. Besides, you may find that the "deterrents" will have a bad effect on your personal health. After all *we* are animals, too.

2. *Use territoriality to your advantage.* Most animals carve out a section of turf as their own; they establish a certain range and tend to stay there. That's a disadvantage because it means an animal will want to stick to where it is no matter how much you shout at it. Nevertheless, once you've coaxed the animal to another territory—either by moving it or by convincing it to move on its own accord—it will tend to stay away. You can do this, for example, by eliminating the animal's source of water.

3. *Remove all sources of food.* Don't leave the pet kibble outside, cover the vegetable garden with netting, and tightly bolt the garbage-can lid down. If there's nothing easily available at your urban restaurant, the animal will go someplace else. This rule applies equally for outdoor critters such as raccoons and indoor ones such as mice. Removing food sounds elementary, but there are some tried-and-true tricks in this encyclopedia that will mean the difference between a few pests and none.

4. Rule number three has a corollary, and it's the motto of all zoos: *Don't feed the animals.* When you give a wild animal a handout, it will expect more. It's the basic response of all creatures: They tend to keep doing the things that get them rewards. Feed the cute squirrel and you'll have a friend for life— or at least for a year. (That's the squirrel's average life expectancy. No wonder they have so many offspring.)

5. *Never handle a dangerous animal.* Animals bite, which is one of their primary defenses. And the consequences of being bitten go well beyond the pain of the immediate wound. Frequently you will be advised to have protective shots for rabies, which can be excruciatingly painful. Here's an additional note of caution to go with this rule: "Dangerous" animals include a far greater number of critters than you might imagine. A reasonably friendly looking raccoon can be a particularly vicious critter when cornered. A mother can stand off a purebred boxer if she's defending her young. Assume that most animals will turn dangerous if they're threatened or attacked.

6. *Block the animal's entrances.* If there are any holes between the outside of your house and the inside, put mesh or brick or some other "discouraging" substance over it. Holes a critter may slip through include heat exhaust pipes, chimneys, pet doors, and ventilation ducts. Unfortunately, deactivating a pet door to keep a raccoon out will also keep your cat in. In other words, some solutions simply create other problems. More about this later.

7. *Erect barrier deterrents.* Fences, netting, Plexiglas covers, even mini-moats all make it difficult or impossible for an animal to get where it wants to go. This doesn't mean that your property will have to look like a fortress, but it helps to at least bar the access to potential food or nesting places.
8. *Distinguish your property from the wild areas.* The more your house resembles a wild place, the more attractive it will be to critters. Painted surfaces, trimmed bushes, well-cut grass, well-maintained entrances, and even exterior lights will create a less "wild" and inviting environment for nature's visitors. These measures help to create a distinction between your dwelling and theirs.

In the pages that follow, we'll show you how to think like the critter you want to control. You'll literally have to get down on your hands and knees to see the world from that animal's point of view. When you do, you'll see the invitation signs, or stay-away signs, that critters see when they approach your home. Although that task can sometimes be frustrating, it's often fun and interesting to outsmart critters. After all, we're supposed to be the most intelligent creatures on the planet. We simply have to use our mental prowess to learn all we can about the animals we're trying to outmaneuver.

These pages contain a great deal of experience and wisdom from people who have succeeded, and even some from people who have failed, in their efforts to control critters. But we had to draw the line on which creatures we thought our readers would need to control. We've included scorpions and alligators, and even zebra mussels (a plague to cottage owners), although we may have left out your own local four-legged nuisance. With a bit of creative tinkering, you should be able to find a solution to all of your critter problems in these pages.

Good luck. And be *patient*. It's helpful to remember that the critters don't actually realize how much they are bugging us.

1

THE CREEPY CRAWLERS

For every human living on Earth, there are 200 million bugs. In fact, the subphylum Insecta comprises 85 percent of all animal species. From an evolutionary perspective, they are the planet's most successful creatures. Along with their cousins the Arachnida, the Insecta accounts for much of the Animal Kingdom's incredible diversity. But from *your* perspective, that's simply a lot of ants, hornets, flies, beetles, mosquitoes, cockroaches, spiders, and ticks to get on your nerves.

It's tempting to think that the world would actually be a better place without bugs and creepy crawlers. Yet, if we destroyed all the insects, we would disrupt the global ecology—and probably wipe out our own species in the process. The backyard fogger that we use to kill the wasps will also kill the predators of the wasp—not to mention what it will do to humans! Fortunately, half of all insect species are predators or parasites of other insects. That means they are helping us control the problem, too.

But short of killing off all insects, how do you solve a local bug problem? How do you keep ants from invading the picnic, mosquitoes from biting your arm, yellow jackets from hovering over the barbecue, cockroaches from taking over the kitchen, and the proverbial fly from landing in your soup? The answer: Each species must be dealt with individually. The solution for cockroaches (and yes, there is a solution) simply won't work for flies. Precautions for scorpions aren't too relevant to blacklegged ticks. And so on down the line. Still, there are four universal truths when it comes to bugs: Laws that are as unchangeable as the insects are pervasive. They are:

1. Bugs will always eat and drink, often voraciously.
2. Bugs will eventually find any food that is not hermetically sealed.
3. You can't eradicate every single insect.
4. Pesticides usually backfire.

Let's spend a minute looking at each of these principles. First, bugs will always eat and drink. In other words, we must have a certain fatalistic attitude toward insects. Whatever you are growing in your garden or yard, whatever you've got stored in your pantry, whatever you've left outside, bugs will find it and eat a piece of it. Expect bugs to be everywhere. They are like uninvited cousins who come to stay more or less permanently, who eat 10 percent of everything that's in your refrigerator, and who never get the hint to do some shopping on their own.

Bugs have skills that even Houdini couldn't match. They'll find anything that isn't sealed airtight. Members of the insect species often have extraordinary senses of smell, which means that the entire outdoors is their domain. Indoors, they will certainly find the box of cereal you've had sitting in your pantry since the previous decade. Even clothes in thin plastic bags are not immune to insect attacks. On the other hand, you can be confident that insects usually won't find their way into refrigerators, Tupperware containers, pickling jars, and items in Ziploc resealable bags. The rest of the universe, however, belongs to them.

You can't eradicate every single insect. There will be some slugs in your garden, beetles in your cucumbers, ants in the grass, flies in your house, and strange ugly things in the basement. You can't anticipate when or where your next bug encounter will occur. Sure, you can take steps to discourage insects, but you can't build a glass box to keep them out. The trick is to deal with them when you see them. You must take steps to minimize their intrusion, and their likelihood of hanging around. Just like the unwanted relatives, if you make them feel too comfortable, they'll never want to leave.

Controlling bugs with pesticides usually backfires. In the short term, pesticides can be effective, but over months or years, insects become immune to them. Insects reproduce quickly, and in great numbers. (Ever notice how there are suddenly flies everywhere in your house on the first warm spring or summer day?) Often insects will mutate in order to resist an insecticide, and those are the insects that live to reproduce and pass on the stronger genes. Scientists are always working on new ways to kill insects, but some environmentalists would say these "advances" are doomed from the start—and that the side effects aren't even worth the effort.

Pesticides have many drawbacks. Simply stated, they are all dangerous, they are hard to apply uniformly, and they're frequently expensive. In your house, the spray insecticides expose you to lots more toxins than the solid bait stations (where you offer one tempting but fatal bug meal) or even the crack-and-crevice foams you can use to seal up your home. Even professionally applied pesticides get into the air, so even if you don't touch them or remain in the house as they're applied, you're still affected. When applied correctly, by professionals, the harmful effects of pesticides may be limited, but they are still present. Pesticides are especially dangerous when used near pregnant or nursing women, infants, people with respiratory illnesses, small children, cats, and dogs (the last three will play and fall on lawns and think nothing of eating grass).

In your yard, the spray you apply to your roses will invisibly sneak into your neighbor's yard and even into the water table. Or it can do even worse. In the early 1980s, a vacationing man golfed for several consecutive days at a country club in Arlington, Virginia. He came down with a headache, vomiting, rash, and high fever. Although he sought medical attention, he died from exposure to a fungicide applied to the grass at the golf course. Pesticides affect animals too. A recent study found that dogs had an increased incidence of certain cancers from exposure to lawn-care pesticides.

The National Coalition Against the Misuse of Pesticides is a Washington, D.C.-based "watchdog" group. It reports that little is known about the nature, frequency, amount, or extent of exposure to the more than 65 million pounds of nonagricultural pesticides applied in and around homes and gardens each year. More than 250,000 North Americans become ill from home pesticide use each year. What effects do they have? If you read a pesticide label, you might notice that 99 percent of the pesticide is "inert ingredients." These ingredients have no effect on bugs. Yet what are they? And what are their effects on humans? Unfortunately, there are very few truly "safe" pesticides—meaning ones that do not have unpleasant side effects. The good news is that there *are* alternatives.

Let's take a systematic look at some of the most annoying critters, and ways we can control them.

ROACHES

When we think of indoor bugs, many of us think of cockroaches first. (The scientific family name is Blattidae.) They have been plaguing humanity since we first dragged home a mastodon leg and left a greasy campfire ring. In fact, cockroaches have been around for almost 3 million years and have a biological system that adjusts to whatever ecological pressures they encounter. There are actually two thousand different species of cockroach, but only four of them make unwanted dinner guests: the German (*Blatella germanica*), the Oriental

(*Blatta orientalis*), the brown-banded (*Supella longipalpa*), and the American (*Periplaneta americana*).

The smallest is the brown-banded at about a half inch, and the largest is the American cockroach (or waterbug) at an inch and a half. They all have flattened bodies and are oval-shaped, with a head that is partly concealed. And, of course, they have those long, wavy antennae that we associate with these little nuisances. Most cockroaches only produce offspring once in their lives (thank goodness), but the German cockroach does it three times. Cockroaches produce 128 to 300 eggs at a time, which are kept safe in a leathery eggcase. The lifespan of the cockroach ranges from six to fifteen months, depending on the species.

Cockroaches are found on every continent on Earth. Even in the two polar regions, they have managed to spread to indoor structures. And we're making things progressively better for them, according to Richard Brenner of the U.S. Department of Agricultural Research in Gainesville, Florida. "As we have gone from living in a Gilligan's Island sort of hut and have evolved to relying so much on structures—not just for where we live and work, but where we store everything we produce—we are creating an ideal environment for arthropods," he says. "We reduce air flow, we create dead air spots, and it's perfect for them."

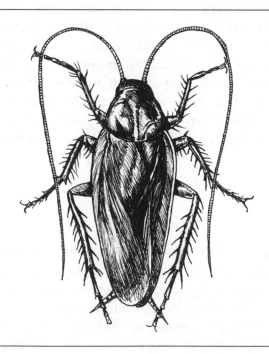

American cockroach (*Periplaneta americana*)

The Creepy Crawlers

Brenner knows what he's talking about. He constructed a model house, mined it with sensors and probes, and then filled it full of cockroaches to see where the bugs would go. The results were impressive, from a scientific point of view. Roaches evidently travel quickly and easily, and they have a knack for getting into the most unlikely places. The Edmonton Oilers hockey team once found a cockroach in a tank of purified water at their stadium!

Roaches are thigmatic, meaning that they like "contact-living." They prefer tight cracks and crevices filled with others of their kind. Most of us have seen at least one of these critters scurry along a baseboard and disappear behind a stove to join its friends. In fact, roaches won't even mate unless their antennae and feet are in contact with something. Offer them a tight, greasy little space and they are perfectly happy. Unfortunately, they also tend to smell, giving off an unpleasant odor that is especially unwelcome in a kitchen.

Roaches walk, crawl, and there's a variety from Florida that can fly and is reportedly making its way north. They are transported in moving boxes, in your morning newspaper, and in the supermarket bags you bring home. Food crates and containers are their favorite means of transportation. So when you move into a pristine new house—certified roach-free—a few cockroaches may have arrived in your moving van or possibly with your first load of groceries. They are the most accomplished hitchhikers in the insect world.

As most people have discovered, roaches are nearly indestructible. Nevertheless, that doesn't stop us from *trying* to destroy them. (And happily, because they are so disgusting, killing them yields little guilt.) The solution I like best, and the one that seems to be most effective, is to buy a lizard that eats roaches. Sure, there's a problem with that solution: A lizard is now running around your house. The good news is that if you choose your lizard carefully, you'll never even see it.

The lizard of choice is a gecko. They are nocturnal, so they tend to do most of their eating at night, when roaches happen to be most active. They also like to hide behind refrigerators during the daytime. There are about forty different species of gecko, and they range in size from several inches to eighteen inches long. They don't make much noise, save for a chomping-on-roaches sound and an occasional bark (yes, *bark*). Many cost less than twenty dollars (don't spend more if you're worried it's heading out the door), and they will live a long time provided you offer them water and bugs. With a gecko housemate, it's quite possible you'll never see another roach in your home.

On the down side, you may have a tough time keeping geckos alive in the colder regions of North America. In that regard, you have to think of them as a pet that needs special care. They will thrive in the hot, humid, southern regions, but they are prone to sickness in cooler climates. It's also not a good idea to use pesticides at the same time as you're trying the gecko method. Poisoned roaches will also poison your gecko. It's also advisable to catch and observe your little helper from time to time. Try to feed it a few crickets to make sure it's getting enough to eat. Although you can't litter-train geckos, you'll find their scat dries quickly and can be vacuumed up.

If you are not inclined to have a lizard roam your house, there are alternatives. Brenner and his associates in Florida have conducted extensive research, and they've found that roaches detest breezes. They prefer stagnant air and the humidity found in dead air spaces in our homes. In their research, air vents were installed on one side of an attic, while the other side remained unvented. After two weeks, the roaches congregated on the unvented side.

Apparently, breezes desiccate roaches, whose survival depends on staying moist. This suggests you should keep your walls and cabinets well ventilated. If you live in a roach-prone zone and you are renovating your house or apartment, make the ventilation of the attics, wall spaces, and cabinets a priority. Why do all of our cabinets have to be flush against the wall, anyway? It's not as if doors keep insects—or even rodents—out. Indeed, keeping your cabinet doors *open* may reduce the bug population.

Brenner is even working with architects to design a home that's less hospitable to bugs. "What we want to do down the road," he suggests, "is develop insect-resistant practices." Already his lab has produced a ventilation device and a new bait that effectively gets rid of roaches. Ridge vents placed along the top of your roof can work hand in hand with soffit vents under the eaves. When air moves along the ridge vent, it creates a negative pressure under the eaves, where air is pulled into the attic. The soffit vents are screened and hinged, so you can open them and place bait in the eaves to kill attic roaches. Remember, these little critters have no problem climbing. You can find them just as easily on the twentieth floor of an apartment building as in the basement.

There's a strategy to controlling any unwanted pest, and Brenner's method makes sense. He calls it his "CIA" approach to insect separation. "In any long-term suppression program, and it doesn't really matter if it is insects or rodents, your chances for long-term success are greatest when your efforts are directed against the stages that are *concentrated, immobile, and accessible.*" You need to look at the CIA factors, just the way Brenner looks at his biological research. Where are cockroaches concentrated? How mobile are they? Are they accessible? Can you get to the population that's causing the harm?

One method for assessing your cockroach problem is the old roach motel: a sticky trap that immobilizes the bugs when they step into it. Sticky traps will let you know where you have roaches. Place them along walls, where the secretive roaches roam, and check them daily, keeping a log of the numbers captured. Sticky traps are obviously good for monitoring but not for controlling, so use them to check on the roach population and change them each month. Every time you rotate them, note how many roaches you catch in a twenty-four-hour period. You can also spot-check the house at night, quietly and gently, with a flashlight. Look for roaches on the kitchen counters, alongside the stove or refrigerator, or near bathroom sinks. Monitoring will give you a good idea of the problem areas. See what you can do to reduce the available food, water, and habitat for these insects as the first step in your battle against roaches.

Once you know where they congregate, force them out, away from their food and water sources. Use caulk or paint to fill in the cracks. You may even have to change some minor design configurations in your house. An elementary school in California, for example, couldn't get a handle on their brown-banded cockroach problem. Then they figured out that the roaches were leaving their eggcases, safe and warm, in the plentiful peg holes drilled for adjustable shelves in closets. After changing the design, officials had removed one of the essentials for an ever-expanding population—a secure nursery. Incidentally, school officials also discovered the roaches were feasting on starch-based watercolors and the cornstarch in an indoor sandbox.

Of course, any treatment for cockroaches begins with better housekeeping. Roaches hate clean homes. The cleaner your place is, the more likely they are to seek out another residence. Leaving dishes unwashed even overnight can be an invitation to roaches. If you need to let a pan soak, make sure that the entire surface is covered with soapy water. Roaches can swim, but soapy water increases the likelihood of their drowning. Throw out old newspapers, especially if they are sitting in a damp area such as a porch. Similarly, old plastic or paper supermarket bags are another roach hazard. Any clutter, in fact, is prime roach habitat.

The next step is to keep all food properly sealed. Invest in large airtight jars for bulk goods such as flour, cereal, sugar, and coffee. Make sure all containers are made of impervious material. Cockroaches will eat cardboard, glue, or anything organic if it's all they can find—especially if it then leads them to more tasty foods inside. At the same time, check those boxes of cookies or other foods that come home from school for cockroach stowaways. And remember that cardboard boxes sometimes used to pack or deliver your groceries may well have traveled—with their cockroaches—from warmer climates. They are guaranteed travel buses for these critters.

Keep "explorer" cockroaches outside your house by tightening up security. Make sure doors and windows are tight fitting, and seal up holes around pipes and wires that enter your house. A constantly refilled caulking gun should be your best friend. Remember, the goal is to discourage their arrival in the first place.

Finally, remove any obvious sources of standing water. Roaches need to drink as much as they eat, and a leaking pipe will guarantee a fresh supply of roaches forever. Insulate pipes that collect condensation. Look for leaks under sinks (particularly in the kitchen and bathroom). And don't let water collect in the saucers beneath your houseplants. Even dripping faucets can be an encouragement to roaches.

Sanitation alone, unfortunately, will not rid your home of roaches. If they are denied food, they will cannibalize. One scientific study managed to keep a captive group of cockroaches alive for two years, even though they were never fed. There's a frightening thought.

When you want to kill the pests, boric acid is an old and still effective remedy, although it's slow and may take a month or two to work on an established population. Dr. Donald G. Cochran, entomologist with Virginia Polytechnic Institute and State University, says the powder is more effective than the solid bait. Boric acid is available at most groceries or drugstores and is deadly to roaches. Its acute toxicity is low, meaning it doesn't kill on contact. The loose boric acid is actually a stomach poison; roaches die of starvation and dehydration. After walking through the boric acid dust, roaches will swallow the substance when they lick themselves during grooming. It then takes three to ten days for the roach to die. There is little or no evidence that roaches have developed any resistance to boric acid.

Here's the preferred technique to use: Sprinkle boric acid lightly—almost like a dusting—along the cracks and crevices in your kitchen, bathroom, or wherever roaches roam. Be careful not to clump it. Use a dust bulb to get the powder into cracks and crevices. If you can, dust the boric acid into crevices and then seal them with caulking. You can also purchase boric acid bait stations, which are effective in damp locations such as basements and bathrooms, or around children and pets.

Boric acid, a naturally occurring mineral, does not evaporate like many pesticides, and it's not absorbed through the skin. Unlike some pesticides, boric acid will also not accumulate in your body. For effectiveness as a deterrent, you'll have to use the kind that's 99 percent pure (look for the Environmental Protection Agency registration number on the label). This will help you avoid many of the mysterious "inert" ingredients found in pesticides we mentioned earlier.

Reports on the effectiveness of boric acid are positive. A Virginia Polytechnic Institute study found the more toxic pesticides to be only 60 percent effective in dealing with roaches. Boric acid had a 90 percent rating. It also tends to have a more lasting effect. A study at the Alabama Agricultural Experiment Station found that the professionally applied pesticide formula Temp reduced a German cockroach population by about 62 percent in a week. But after another twelve weeks, the population was right back where it started.

One compound introduced in recent years is Drion, which contains silica aerogel. It actually kills roaches by drying them out. Both boric acid and Drion, however, must be used carefully, *especially around children and pets*. Even a teaspoon of boric acid can potentially kill a child. Similarly, it's important not to inhale the stuff as you apply it. Always wear a mask, gloves, long sleeves, and long pants.

While some of the commercial bait stations contain a growth regulator, you can purchase one on your own. A product called Gencor 99% prevents roaches from reproducing. The spray is applied around areas where you see roaches, and it lasts up to six months. Gencor 99% has a good reputation for effectiveness.

If *any* commercial pesticides are unthinkable to you, try a home brew. Fill a wide-rimmed jar with beer, honey, and part of a banana or its peel. Wrap the outside of the jar with masking tape so the roaches can climb in. Then coat the inside rim with petroleum jelly so they can't climb out. By the next morning, you'll have more roaches than you could possibly want—or ever knew you had in the house.

Dr. Cochran still recommends bait traps, such as Combat or one of the Raid traps. "They're in a contained unit so pets and kids can't get to them," he notes. "And you control the dose yourself." These poisons kill in three to five days, and the package suggests the best spots for placement. Target the inconspicuous places where roaches like to travel: behind appliances and in the space near the joint between the floor and wall. You may be able to push these traps into the wall space through the hole around the drainpipe of your kitchen sink.

If you have an out of control roach infestation, you might be inclined to use drastic remedies. But first, how do you know if your roach problem *is* out of control? When you enter your kitchen or bathroom during the night, turn on the lights, and see roaches on every surface, that's a serious problem. In this case, you'll see scores of them—not just one or two. If you see the nocturnal German roach, the most common household pest, during the day, that's also a sign of infestation. Catching thirty to forty cockroaches in a sticky trap in a twenty-four-hour period would certainly count as an infestation, too.

Skip the indoor foggers and those automatic bug sprays. The toxicant gets everywhere and on everything. That means on the furniture, on your bed linens, on the dishes, and on your clothes. (Here's an example of the solution becoming a greater problem than the problem itself.)

You may be wondering why you have to do anything at all about roaches. After all, what's the problem or the harm? Our expert, Dr. Cochran, points out that there is solid evidence that roaches are capable of transferring bacteriological diseases. For instance, your average house roach is not averse to feeding on the soiled litter in the cat's litter box and then tasting a bit of the cake you left under plastic wrap in the kitchen. It's a disgusting thought, but it's painfully true. Research also indicates that allergies caused by cockroaches are the second most common problem among asthmatics, affecting 10 to 15 million North Americans.

On the subject of what to avoid, you might want to add ultrasound devices to the list. These instruments emit high-pitched sounds we cannot hear, but which other animals can. Unfortunately, all it seems to achieve with cockroaches is an avoidance pattern. Initially, it forces the bugs to move around and change their feeding patterns, which might appear to the homeowner to be a positive change. The roaches don't die or move out, however, they simply vary their routine and take up residence in "sound-shadowed areas." In other words, they just figure out which room, wall, or refrigerator blocks out the noise, and they spend more time in that area.

Finally, it's worth noting that not everyone dislikes cockroaches and wants to get rid of them. Exterminators, for example, are delighted to see them propagate. But there are also amateur and professional cockroach racers. You can find these people and their trained critters at the annual Great American Bug Race, with their champions all lined up antenna to antenna. Amusing? Maybe. At least until you have to deal with a race in your own kitchen in the middle of the night.

ANTS

Getting rid of ants is no picnic, if you'll pardon the pun. Ants seem to view the *individual* as expendable, but the survival of the *species* as everything. Smack one with your kitchen sponge, and instead of scurrying away, they'll send out the reinforcements. Another 250 ant soldiers march under the door, up the cabinets, across the wall, and down into the sink to do battle with you.

Ants are hard to discourage. Clean the sink, turn your back, and they'll form a line into the garbage. Take out the garbage, and you'll find them swarming over a single Cheerio under the kitchen table. Scour the kitchen, and they show up in the bathroom. Their strategy is to use sheer numbers to wear you down until you surrender and declare your home a national indoor picnic site. There are some effective control strategies, though.

There are hundreds of species of ants (genus *Hymenoptera*), but only about twenty-five species commonly infest homes. Pest ants are usually divided into two groups based on their typical nesting preferences, either wall-nesting or ground-nesting. They all tend to be highly organized insects with a strong social bond. Each "colony" may have several queens, as well as a host of males (drones) who do nothing but fertilize the queen. Meanwhile, the worker ants endlessly gather and store food for the colony.

Many species nest in the ground, but they are often content in any protected cavity. Some are meat-eaters, some are scavengers and will eat anything, and others prefer plants. If they move indoors, that's bad news for us: It means that just about anything foodlike is in their diet. Almost the worst thing you can do against these household invaders is to use conventional pesticides, which make the colony split up into multiple colonies. They seem to be saying, we'll divide and then we'll conquer you. Because they have so many queens, they can make lots of colonies—all over your house.

Most ants will bite if you disturb them, and they can get off the ground to do this. The queens and males actually have wings and are capable of flying. (The wingless ants you often see are the thousands of workers.) Many ants will also sting you if they are threatened. A few of them can even send off a putrid smell from their anus to detract intruders. Ant bites can be itchy and painful, although the worst reaction usually comes from a fire ant (*Solenopsis* spp.). These little devils will bite and sting you, and they inject a type of formic acid into your blood that may cause an intense reaction in some victims. Fortunately, fire ants tend to be found outdoors, and you may be lucky enough to see their swarming mounds before you mistakenly step on one.

Boric acid, once again, is the best method to cope with ants. They aren't going to walk through it and then lick it off like cockroaches do; you have to slip it to them in their food. The worker ants eat the poison bait, and then take some back to the nest to feed the queen and the young. The liquid forms of boric acid, marketed under the name Drax, are most effective for ants, and you can buy the bait from a number of mail-order catalogs. Or you can simply make your own.

Here's the recipe: Mint-apple jelly is one of the more tempting baits for pharaoh ants (*Monomorium pharaonis*). Mix one teaspoon of 99 percent pure boric acid into one-third cup of the jelly. Place the bait along a piece of tape, in bottle caps, or simply on the floor or counter where you have seen ant activity. You

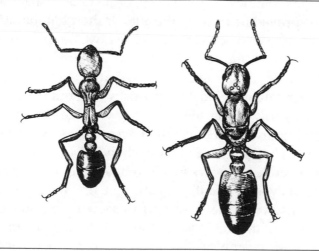

Pharaoh ants (*Monomorium pharaonis*)

don't need to block their trails with bait; each daub treats about twenty-five feet. Check the bait (commercial or homemade) every three to five days, and replace the dried-out bait. If you're frugal, you can simply refresh the bait with water. You can also use corn syrup as an attractor.

The hard part about this treatment is watching the ants feed at the bait stations while you stand by and seem to do nothing. The good news is that they will crawl away and die. It takes one to two months to clear away the ants using boric acid, but it's certainly an effective remedy.

Some varieties, such as carpenter ants (*Camponotus* spp.), like to live indoors. Other ants that you'll see around your house are visitors from outside. They are simply sampling your food supplies while they keep their nests out in your yard. With these creatures, you need to find where they get in and block the entrance. If you are so inclined, you can kill them with poison, but not a bait, which would just encourage more deadbeat ants to come in and eat your food. Try sprinkling boric acid powder again, this time around the areas of activity.

A product called pyrethrum made from pyrethrum flowers, a member of the chrysanthemum family, works well, too. Just because it is botanical, does not mean it isn't deadly; it's 25 percent more acutely toxic than boric acid. Dust the areas where you find ants. Pyrethrum kills a variety of household pests and breaks down well. A caveat about using it though—unfortunately, the farmers who are clearing land to cultivate pyrethrum are destroying sensitive mountain gorilla habitats in Rwanda, near the camp that Dian Fossey established to study the species. Perhaps other sources will become available soon.

Another possibility is household cleaner. Interestingly, it makes a terrific spray that's effective against ants as well as other bugs. Household cleaners also work faster than pesticides. Windex, Formula 409, or just about any spray cleaner will do the trick when sprayed directly on the ants or their habitat. Kills 'em dead, as the slogan says. Even less toxic than aerosol sprays, but just as effective against ants, is a mixture of water and soap. Next time you are finished with a bottle of Windex, fill it with soapy water and label the bottle BUG SPRAY. Your homemade mix or some commercial insecticidal soap works well but is harmless to humans.

While cleaners work well on the ants that you see today, the best strategy against future visiting ants is to block their access. So spare a few and watch to see where they exit. They generally follow chemical trails they have left in and out of the house. A caulk gun works well to block the holes. Ants are persistent insects, so do not put away the caulk gun after the first incident; you'll soon use it again. You should also take a look around to see where you can clean up food sources. Small ants can climb up loosely closed screw-top jars to die sweet deaths in honey or liqueur, for example. They're attracted to open pet food, spills, and garbage of any kind.

The Creepy Crawlers

Ants in the yard are less of a problem. Nonetheless, if you notice a population explosion, look for food sources: fat spilled under your grill, exposed garbage, partially burned refuse, a compost heap that's not composting well, or the infamous pet food. Sometimes ants may be attracted to feed on the sweet "honeydew" produced by scales (very small tree and shrub bugs) and aphids (winged or wingless) that are attacking plants in your garden. Deal with the aphids and scales, and you have eliminated the food source. Another common problem is rotting wood, which is a favorite home for carpenter ants. These giants of the ant family may live in an old stump in your backyard, but they will happily venture inside to poke about for food or even to nest.

You may be able to drive out the ants with water after you have eliminated the food sources. Flood each ant hole with water from a hose by pressing the nozzle against the earth, or pour soapy water down the hole. Deterrents such as boric acid or pyrethrum used outside will kill a wide range of insects, so be careful if you use it in open spaces. Also remember to wear protective clothing and rubber gloves when you are using these toxic materials. And don't forget to wash your clothes and take a shower afterwards.

Ants are a favorite meal of many backyard birds. Woodpeckers, for example, are voracious eaters of these insects. You may also notice a large member of this family, the flicker, who is more of a ground feeder. It will find and devour enormous numbers of ants around your house. Keep your cat indoors, and the flicker will be your well-trained ant exterminator.

Keep in mind that all ants are not bad. There's a beneficial side to them as well. They are the sworn enemy of termites, and they eat fly larvae, too. So if you have an ant infestation in your house, attack only the problem. Don't eradicate all of the ants in the neighborhood, or you might invite a plague of other insects.

TERMITES

If you mention the word "termites" (of the order Isoptera) to most people, they have a vision of their house crumbling around them after the critters have eaten through the foundation. These little soft-bodied and pale-colored insects are social creatures, just like the ants. Where there is one, unfortunately, there are many. Again, like ants, they live in colonies—all too frequently indoors. There are almost 2,100 species in the world; about forty of these live in North America. Drywood termites and rotten-wood termites are two of the most annoying varieties you'll encounter.

Their main food is wood, as is their habitat, although you'll also find them living underground. There are queens, soldiers, workers, and reproductives, all

KEEPING YOUR HOME TERMITE-FREE

- Store your firewood and other potential termite foods such as lumber away from the building. Do not stack wood against the exterior of a house, for example. That is simply inviting them inside. And remove old stumps that are near your house. Think of wood as termite bait and handle it accordingly.
- Keep your basement and the area outside your basement as dry as possible. Moisture is essential for the survival of termites. Make sure that the drain spouts work properly and that bushes don't come in contact with your house. It's also important to inspect your house regularly, particularly the supporting beams and other critical building areas.
- Keep your house well-painted. Cracked and peeling paint offers termites a way indoors. The areas adjacent to the ground are critical. Exposed wood supports are major offenders. Always coat your wooden building materials with paints, stains, or resins that seal the wood and discourage creepy crawlers.

assigned their various tasks. One of those tasks, of course, is to eat the wooden foundations of houses, or to munch on telephone poles, fence posts, or even your furniture. They tend to eat from the inside, which means that the wood then crumbles and collapses. It's not hard to see why they create fear in the minds of homeowners. The only beneficial aspect of termites is that they help to decompose fallen trees in the woods. Hardly a reason to adore them, however.

But what if you *do* have a termite problem? In spite of what exterminators will tell you, pesticides are not the only option for dealing with these critters. Nematodes are a sensible alternative to pesticides. These tiny, wormlike creatures are deadly parasites to termites. And unlike pesticides, nematodes will not harm your children or pets—or anything other than termites. Unfortunately, finding a company that uses nematodes may not be that easy, although your local environmental groups can often give you a lead. Nematodes *can* be ordered through the mail from a number of organic gardening supply companies.

There are two other potential solutions that you might consider. The first is rather dramatic but effective: Freezing your house with liquid nitrogen. This lowers the temperature in the walls to 20 degrees below zero—a frosty temperature that spells death for termites. (That's why there are supposedly no termites in Alaska.) It may also kill all the roaches and ants. Make sure your pipes have been drained beforehand or the water in them will freeze, expand, and possibly burst the pipes.

Alternatively, here's another dramatic solution: heating your house to 120 degrees Fahrenheit by blowing hot air through it will rid your house of termites. This technique will rapidly kill termites and any other insects that happen to be around. As a precaution, remove all candles before you pursue this option. And cans of Coleman fuel!

BEETLES AND GRUBS

The "numbers" on these little critters are truly staggering. One-fifth of all the total plant and animal species in the world are beetles (order Coleoptera). North America is home to about 17,000 of the more than 350,000 known species of them. Bear in mind that they go through four life stages—egg, larva, pupa, and adult—and you can see that there are a spectacular number of beetles in your house or yard at any moment.

Most beetles are plant eaters, which you'll know if you have ever tried to maintain a flower or vegetable garden. They are admittedly hard to control, even with pesticides. This is especially true of Japanese beetles (*Popillia japonica*), which feed on more than two hundred varieties of plants—most often the ones that seem hardest to cultivate. As adult beetles, they eat your roses; as grubs (the larvae), they create patches of dying grass as they feed on the roots. Through the course of their life cycle they first attack the roots of plants in your garden, then the flowers and fruits, and finally the leaves as they mature.

The most destructive beetles are the members of the snout family (Curculionidae). There are more than six hundred members of this nasty group in North America, including the European elm bark beetle. It's been responsible for the plague of Dutch elm disease. The other infamous snout beetle is the boll weevil, which ruins cotton crops. These beetles have hard shells, with a long snout, and are anywhere from one-tenth to one-and-one-half inches long. They are prevalent right across North America.

Here are some simple steps you can take to help eradicate these crawling pests. The first tip is to avoid beetle traps. These commercial traps use a scent that *attracts* the insects. In this instance, that's a mistake. If you own a large piece of land, however, it's okay to place a trap in a distant corner of the yard where you may not draw beetles to your prized plants. But placing one near your garden is just inviting disaster.

Next, if you want to try to vacuum Japanese beetles off your plants, or hand-pick them, the morning is the best time. That's when they are most slothful. Squish them between your fingers, under your heels, or whatever method you prefer. If you're squeamish, just drop them in a bucket of soapy water.

The third piece of advice is to pay attention to the life cycle of beetles. Japanese beetles lay their eggs in July and August; the less you water your lawn during those months, the fewer beetles you'll have. Beetles feed until October and November, when they move below the frost line and wait out winter. By March, they're eating again. You can spot grubs if you peel back a layer of sod and peek beneath. If you see ten or more per square foot, you have a problem.

Try using *Bacillus popillae*, also known as milky spore. It often kills beetles in their grub stage, when they live in the dirt under your lawn. Milky spore effectively fights soil-living grubs of the Japanese beetle, rose chafer (*Macrodactylus subspinosus*), oriental beetle (*Exomala orientalis*), and some May and June beetles (*Phyllophaga* spp.). It is sold under the name Doom or Grub Attack at garden centers, or you can order it from a variety of organic supply catalogs. Apply it by sprinkling the dust onto the lawn and then watering it so it sinks in. Treating for lawn grubs also stops any grub predators like moles or skunks from tearing up your lawn in search of a meal. (You'll also find that raccoons like to roll back the turf in search of these little snacks.) The spore kills the grubs and remains effective in the soil for twenty years. Plus, when the grubs die, the decomposing bodies release more of the spore into the soil. It is harmless to humans and pets, and it won't hurt your earthworms either. Unfortunately, milky spore takes up to three years to get established, so you certainly won't see the results immediately.

A quicker killer can be found in parasitic nematodes of the strain Hh, also available from organic gardening suppliers. These little killers enter the host body and release bacteria that will terminate the grub in only forty-eight hours. The nematodes feed on the grub's remains and then reproduce.

The most effective grub-killing award goes to some Colorado State entomologists. They used soil aerator sandals. In the early and late summer when the grubs were feeding heavily near the surface, the entomologists put on their spiky sandals and methodically walked over a lawn several times to get two nail insertions per square inch. It took a lot of walking, but it did the trick.

SLUGS

A slug is actually a kind of land snail anywhere from one to five inches long, but without a shell. (That means it's part of the Mollusk family.) You will recognize it by the soft, flabby, fleshy, and slimy body, and, in some species, by its spots. Slugs vary in color from white or gray, through yellow, brown, or black. They like to eat the leaves and stems of a variety of plants and shrubs, but they also munch on some soft fruits as well. You'll see their trail; they leave large, irregular-shaped holes in everything they attack. Their telltale slime may also be visible.

The most common variety is the great slug (*Limax maximus*), which is one of the longer varieties at three to five inches. It likes both the city and the country and is the true pest of all gardeners. During the day, it hides in cool, damp, and usually dark places, and then emerges at night to feed.

One of the myths about these garden pests is that marigolds will keep many of them away. That's not true. In fact, slugs will kill your marigolds! There is another tried-and-true slug-eliminating trick, though. Take a few jars and bury them up to their rims in the soil, then fill them with beer (cheap domestic brands work as well as the imported). The next morning, your jars will be full of stale beer and dead slugs. Dump the brew onto the compost heap and set the jars out every night until you're convinced the slugs are all gone.

If you want to go high-tech, you can buy the commercial equivalent of a jar of beer that comes with a roof and a removable cup. One brand name is Garden Sentry (there are other similar types available from garden suppliers). It too gets buried up to its rim in the garden and is also filled with beer or commercial bait to lure the slugs into it. Once inside, the slugs will drown, since they can't slither back up the steep sides. To get rid of dead slugs or to refresh the bait, simply lift out the sieve-like, removable cup and leave the device in the ground. The roof keeps the bait from being diluted, but it may catch on your lawn mower's blades if you're not attentive.

To help prevent slugs, you should get out your rake and give the garden a thorough thrashing a few times each year. This will help to destroy the slug eggs that are nesting in your soil. Hostas are one variety of flower that seems to really attract these pests, but there are certain varieties that are slug resistant. Try the Sagae, Elegans, Sum and Substance, or Blue Angel varieties, for example.

Some gardeners prefer to handpick slugs at night or in the early morning. You can also trap them with flowerpots turned upside down, or you can attract them with damp newspapers or boards laid out on the lawn and garden. Garden experts also recommend diatomaceous earth. This powdery natural material composed of fossilized algae seems to discourage slugs. Take a few handfuls and spread it around the base of your plants (always wear a dust mask when you're working with diatomaceous earth). Another suggestion from "earth-friendly" gardeners is to use iron phosphate bait. It comes in little granules that you sprinkle in a band around the affected plants, or in damp places where the slugs may be hiding. Apparently it causes them to stop feeding, and they die within a few days.

Another suggestion is simply to pour salt on them. They usually shrivel and die right before your eyes, which provides a measure of revenge if they've just destroyed some of your prized flora. If you get desperate, you can also try metadehyde—which appears to be a surefire slug killer. This stuff was originally used as a solid fuel for camp cook stoves. (It does, however, have the potential to do serious kidney damage in children and pets if ingested in even small doses.) Copper edging around plants seems to work for some gardeners. A solution of ammonia and water can also be an effective spray. Ortho's Bug Geta Snail/Slug bait also gets good results. And one final note about slugs: It seems that they are a favorite meal of geese and skunks. Each of *them*, of course, brings its own brand of critter problem.

Other Yard and Garden Pests

One common technique is to attack garden bugs with a hand-held vacuum cleaner. Stalking your garden and swooping bugs into this contraption offers a great deal of satisfaction. And it's very effective. Vacuum cleaners work best against leaf-eating insects such as aphids and beetles. Empty the bag into soapy water to kill any insects that survived being sucked into the machine.

Unscented sticky traps also work well. Yellow ones mainly attract aphids, whiteflies, leafhoppers, black flies, moths, and gnats. White attracts tarnished plant bugs. You can clean the sticky trap's surface with vegetable oil and, to reinvigorate it, brush on a liquid sticky coating such as Tangle Trap.

You can also recruit beneficial bugs to do your dirty work. Predator insects include ladybugs, lacewings, praying mantises, and spined soldier bugs. These insects are available from organic gardening suppliers. Some other powerful organic weapons to use against yard pests include the nematodes mentioned earlier and selected forms of bacteria. One bacteria, *Bacillus thuringiensis* or Bt, controls an entire plague of insects: cabbage loopers, cabbageworms, diamondback moths, tomato fruitworms, grape leaf folders, and gypsy moths.

SCORPIONS

Most people have a general idea of what scorpions look like: scary crawling creatures with long curved tails, pincers at the front, and stingers on the tips. We've seen them in James Bond movies, or perhaps at the zoo. And we know that they are relatively small but thoroughly terrifying.

Scorpions are actually related to spiders and mites (the Arachnid family), and they are found throughout most of the world. Many are a ghoulish yellow color, and some can move at the speed of a bullet. Some scorpions are tiny:

only a half inch long. Others are as large as eight inches. All of them have four pairs of legs, plus a pair of pincers (called pedipalps) that make them look a bit like a lobster.

Estimates vary, but there may be as many as fifteen hundred species of this creature. In the state of Arizona alone there are at least thirty varieties. They are generally secretive, hiding during the day under rocks or in tight secluded spaces. At night they emerge to eat spiders, insects, and occasionally each other. In terms of their lifespan, they can live to be more than six years old. They can live for months, according to scientists, on water alone. That's why it's a good idea to check for them near water sources in your house.

Only one kind of North American scorpion has venom considered potentially lethal to humans, and that's the Arizona bark scorpion found in the Southwest (*Centruroides exilicauda*). It is one to four inches long (usually averaging about two inches), yellowish in color, slender, with eight legs, two pincers at the front, and a curved tail that it carries over its body. Its bite is potentially lethal, although it is not aggressive by nature. You'll sometimes see it in the desert, where it lives under rocks, logs, or other debris. Backpackers, for example, have to be careful because scorpions are known to crawl into sleeping bags and boots to get warm. But a scorpion is also a good climber, so it is equally happy in trees and shrubs, where it will easily find insects to eat.

Scorpions are solitary, nocturnal animals, and they only come together for mating. It's hard for us to tell the difference between males and females, but obviously they can. After a complex mating dance in which they entwine tails and scoot around, the male fertilizes the female's eggs, which she carries internally. If the male doesn't head off quickly, the female may kill and eat him after they mate. (This interesting habit has led to some of the mythology and lore about the creature.) Depending on the species, the young are born after several months to a year. They rest on their mother's back for days, until they molt and are ready to hunt on their own. Scorpions may live for several years in the wild (and yes, some people keep them as pets).

Although they can spend long periods of time without food, they do like to hunt. Scorpions eat a wide variety of vertebrates and invertebrates—everything from insects to mice. They wait in hiding and surprise their prey, but they don't chase it down. Scorpions grab with their pinching front appendages and then sting and paralyze the prey. Eating takes a long time for scorpions because they don't chew their food. Instead, they ooze digestive juices over the victim and then suck up the predigested meal. (It's rather gross—but fascinating.) In turn, scorpions are eaten by birds, lizards, mammals, insects, spiders, and other scorpions, which disable the stinger before liquefying their prey. Despite popular folklore, scorpions do not commit suicide by stinging themselves.

All scorpions bite, but only a few of them—including the North American bark variety—can be fatal to humans. Still, the bites do cause pain and numbness and can bring on local paralysis, fever, and general flu-like symptoms for a few days. Most scorpion stings are accompanied by swelling, tenderness, and warmth in the area of the bite. It's a good idea to call a poison control center to check if the symptoms are benign. Other symptoms, such as muscle spasms, may occur and should be treated medically. Otherwise, home treatment might include the use of ice packs, antihistamines and analgesic ointments, and bed rest.

Most of the time you'll find scorpions outdoors, in washes, rocky areas, under rocks, or in burrows; however, the one scorpion that likes the indoors more than other members of the species is the nasty bark scorpion we've described above. You'll find it in Arizona and New Mexico, and reportedly in California on the west side of the Colorado River. It gets its name from a preference for old wood; it likes log piles and rotting walls. And it has a special fondness for old, dead palm trees, where it crawls under the bark.

The bark scorpion can cause death, often due to respiratory paralysis and other complications, usually within one and one-half to forty-two hours. Children are especially at risk. The sting of the bark scorpion is immediately painful, sometimes causing numbness or tingling in the area around the sting. Swelling, however, is rare. Other symptoms to watch for, especially in children, are restlessness and random movements of the head, neck, or eyes. Adults will notice increased heart and breathing rates, and have difficulty catching their breath. Blurred vision and muscle twitching may also occur. If you notice any of these symptoms, especially in children, take immediate action.

A low fatality rate for bark scorpion bites has been achieved, mainly because of quick responses. Every year, some people do die from scorpion bites, but very few in Arizona where antivenin for the bark scorpion is readily available and most medical centers know how to treat the problem.

First aid for a bark scorpion bite starts with keeping the victim calm (that minimizes the absorption of the venom). Wash the area immediately with soap and water and apply a pressure dressing and cold pack to the sting. At the hospital, doctors may administer oxygen to assist with your breathing and prescribe pain relievers. If a victim is not responding or is having a severe reaction (especially if it's a child), then most doctors will recommend an antivenin.

It's unfortunate that the bark scorpion has discovered the great indoors. In fact, it almost seems they prefer human habitats to those provided by nature. As Carl Olson, an entomologist at the University of Arizona, suggests, "They're adapting very nicely to the habitat we give them." How do they get into houses? According to Olson: "There always seem to be little cracks hither and yon.

COPING WITH VENOMOUS BITES

- Remember that 90 percent of these bites occur on the hand. That's the first place to look for possible signs of an attack.
- If it's *safe* and possible, capture or kill the scorpion and put it in a container. Even if it appears to be dead, however, don't handle it with your bare hands. Instead, scoop it up with a piece of paper and take it with you when you go for treatment. Identification can make it much easier for the doctors to effectively treat the bite.
- It's a good idea to get medical care even if the sting does not feel severe. Quick medical attention, especially for bark scorpion bites, is essential.
- If you cannot get to a medical facility quickly, call a poison control hotline for advice. Dial 911 and an operator will connect you.
- Do not attempt "home treatments" other than keeping the victim calm and administering an ice pack.
- If the victim is a child, is elderly, or is in ill health, consider any bite to be a medical emergency. Most deaths from a bark scorpion have involved children.

They're not hitchhiking on animals or on you. They just walk in—they're very thin—and they seem to get through all kinds of spaces."

So what's the strategy for stopping scorpions? The first line of defense is to cut down on the numbers that come in. Once again, arm yourself with a caulk gun. Close up any door openings, or places where screens are torn, especially in windows. (Remember that scorpions can climb right up walls.) Because the spaces they can crawl through are so small, it may be impossible to block up every avenue of approach; however, this is the first place to start. Attach weather stripping along door seams and add screening to any vents.

It's also a good idea to keep corners and out-of-the-way places in your house well dusted and vacuumed. Leaving them full of cobwebs just invites scorpions to set up homes. Remember that common pesticides usually don't work on scorpions. (If you insist on using them, contact a professional exterminator. They can do a black light test to check for the presence of scorpions because they actually glow in the dark.) And be sure not to walk around the house at night with bare feet!

The next step is to evict the scorpions that have already entered the house. Where to look? Here's Olson's advice: "I've found them all over, but one of the most common places is where there's water, because they need to drink, too. So when there's a free source of water, like in the kitchen or bathroom, they get in there and they can't get out, because they can't climb the porcelain." Don't leave standing water, and remember to wipe the shower dry.

Olson reminds us not only to look for scorpions when we're on our Sunday-afternoon pest purge, but also anytime we're in the kitchen or bathroom. "People get into the shower, step on them, and get stung because they didn't look," he explains. Scorpions can also be found on walls. "Most of the time, you have a chance of seeing them, because they cling to the walls; they are climbers." When Olson finds a bark scorpion in his tub or on the wall, he simply captures and releases it. "I just put a jar over them to catch them." Yet, as the warning goes, don't try that at home, folks.

A third line of defense involves removing scorpions from your outdoor living areas. Start by going around your house at night with a red-covered light, which scorpions cannot see. Find the scorpions that are outside the house, and move them elsewhere (get a professional to do this—some are fast enough to scoop scorpions up into a jar). Although a few fearless amateurs may try this feat, not every reader will feel brave enough to attempt it.

You can then try to push their habitat farther away from the building. Start by removing big rocks; this is a favorite place for them to hide. Get rid of dead wood and trees and brush close to the house. The idea is to create an inhospitable zone between your house and their dwelling areas, so the scorpions are less likely to want to venture across that gap.

Dealing with scorpions—as with many other wild creatures—is also a question of attitude, says Olson. "People come to the desert, but they don't want to live in the desert. They won't accept that this is part of desert living." Instead, he says, "You just have to be careful. When you go hiking in the desert, you look where you put your feet so you don't step on a rattlesnake." It's the same with scorpions. "It's a matter of opening your eyes and being aware of your surroundings," not of relying on chemicals to "surround you and protect you from these nasty, vile things that are out there."

Olson objects to the use of chemicals as a remedy. In fact, he says some pest control companies play on people's fear of scorpions in an effort to get them to use chemical pest control products, which, in his opinion, "don't do any good." He explains: "You can put down the chemicals, and the scorpions will walk through it, and they'll still come into your house." Eventually, the poisons will kill the scorpions, but not before they've had a chance to explore your house. In other words, they'll still cause you to shriek at least once.

And it doesn't look like these encounters are going to stop anytime soon. True, scorpions are not at epidemic proportions like some other pests. And they are certainly not prolific like cockroaches or aphids. But as humans encroach farther into the territory of animals, we should expect to encounter more nonhuman neighbors. That's especially true of scorpions.

Scorpions bear quite a few young—maybe thirty a year. And many of those young survive because in an urbanized setting, fewer and fewer of the scorpion's natural enemies survive. At the same time, our living areas provide scorpions with places to hide and moisture to drink, while our houses attract other bugs, such as crickets, which they like to eat. "Everything we do that's nice for us is nice for the scorpions," says Olson. "So we make the problem."

Here's one final suggestion. For detailed instructions on making your home scorpion resistant, check out *Goodbye Scorpion; Farewell, Black Widow Spider* by Dr. David Hawkins. You'll find some good advice about how to live with—and without—this critter.

SPIDERS

One of the most feared members of the insect family is the black widow spider (its technical name is *Latrodectus mactans*). It is one of several "widow" species of spiders and inhabits every state except Alaska. You'll find it most often, however, in the south and west. This creature is very elusive and shy, preferring to hang out in attics and under floorboards. It catches its prey—other spiders and insects—in its web, which has an erratic shape. In fact, that's one way to identify this spider. Most spiders have uniform patterns to their webs, while a black widow spins a more haphazard web.

As befits its reclusive nature, the black widow spider only attacks if provoked. As with other spiders, if its mate dies, it might eat it, and it might even kill its mate in order to eat it if other food is not available. Some male black widows mate with more than one female, however, and most die of their own accord after mating. Females, meanwhile, are known to live up to three years.

The female and male black widows are shiny black and have the same shape, but they differ in important ways. The female, with a thirty- to forty-millimeter (about one and a half inches) leg span, is larger and it has a red hourglass marking on its belly. The male is smaller, with a sixteen- to twenty-millimeter (three-quarter inch) leg span, and it has red and white marks on its belly. More important, the bite of the female is venomous to humans, while the bite of the male is not. (This is the source of much of the popular lore about these creatures.) It is not clear if the venom of the male is less potent, is present in less quantity, or if the shorter fangs of the male do not inject as much venom with a bite. An immature female's bite is similar to that of a male.

Contrary to popular belief, the bite of a black widow spider is not a death sentence. Death is rare, with a rate estimated from less than 1 percent to 4–6 percent. Some people are not even aware they have been bitten. Others experience a sharp pain, as if they've been pricked with a needle; this is often followed by a dull, sometimes numbing, sensation around the bitten area. Bites typically occur on the hands and feet; however, some unhappy visitors to an outhouse have also been bitten in the buttocks or genitalia.

When it bites, the black widow injects a toxin into its prey. The usual victim is another arthropod, such as an insect or spider, which provides food for the black widow. The toxin paralyzes such tiny creatures; in humans, it causes a variety of symptoms, including severe muscle pain in the abdomen, chest wall, and other parts of the body. It can also cause restlessness, anxiety, sweating, headaches, breathing problems, nausea, and vomiting.

A black widow spider bite should be treated as a medical emergency. You should take the victim to an urgent care facility or the emergency room of a hospital, especially if the person is young, elderly, or in poor health. If the reaction to the bite is a mild one, it will resolve on its own, but if the reaction is severe, things can get complicated. Medication usually counters the effects of the toxin. Antivenin is often available, but it should be used carefully; it might bring on further reactions. In severe cases, and particularly for patients with heart conditions or high blood pressure, the victim will need to be hospitalized.

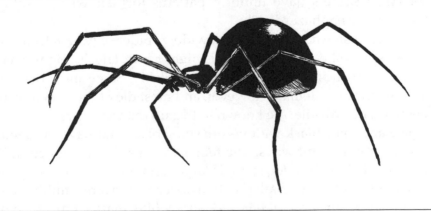

Black widow spider (*Latrodectus mactans*)

The Creepy Crawlers

The brown recluse spider (*Loxosceles reclusa*) is less well known than the black widow and its U.S. range is more limited: generally from the East Coast westward to Arizona and Wyoming. It is medium size, about ⅜ inch long, is tan or brown with a violin-shaped marking near its head, and sports six eyes. It likes a dark and undisturbed habitat, such as woodpiles and storage closets. Most bites occur when a person reaches into a brown recluse's space, or puts on clothing, especially little-used items, in which a spider is resting.

The bite of a brown recluse spider may feel like an intense sting or it may even be painless. Symptoms appear in three to eight hours, when the bite site becomes red, swollen, and tender. The toxin injected can cause death of the tissue around the bite, and the sore can become enlarged. It will slowly heal, but it may take six to eight weeks and the lesion can leave a prominent scar. One study indicates that tissue death occurs in less than half of documented brown recluse bites.

Seek immediate medical attention for a brown recluse spider bite. Although the primary symptom is a rash or deterioration of skin at the site, the problem can become serious. Some bites may lead to nausea, anemia, weakening of the body, or even a coma. Deaths, although extremely rare, have certainly been documented. To treat the bite, use ice to slow the action of the toxin, cut out the wound (although this treatment has recently been questioned), and treat with antibiotics. An antivenin may be effective, but it's often not widely available.

Because many benign spider bite symptoms resemble the bite of the brown recluse, it's a good idea once again to save the captured or killed spider for identification. This will help the doctor to form a proper diagnosis and offer the right treatment. There is good news about spiders, though. Some of them can give you a painful bite, but they are also very helpful indoor bug-removers. Spiders will seriously reduce the number of other insects living in your home. It's a trade-off: Reducing the spider population may make other indoor pests more annoying for you.

CHIGGER MITES

The skin affliction known as "chiggers" is caused by the tiniest of critters—the chigger mite (*Trombicula irritans*) in its larval stage. The larvae are barely visible to the naked eye measuring about ¹⁄₁₆ inch in length and may be red, orange, or yellow. Fortunately, chigger mites do not transmit disease, and the irritation they cause is not contagious.

These little creatures are found throughout most of North America. They prefer moist habitats such as woodlots, pastures, roadside ditches, and other areas that are grassy, weedy, woody, or marshy. They perch on grass stems, leaves, and other shrubbery, and when animals, including humans, brush against the plants where they lie in wait, the chiggers hop aboard. Once attached to a host, they search for spots where clothes fit tightly or the skin is tender, such as at the waistline, the groin, or the back of the knees. Contrary to popular myth, they do not burrow under the skin—they simply insert their mouthparts into the skin and start to feed. Unlike their relative the tick, chigger mites do not make a meal of a host's blood. The larval chigger mites inject a salivary secretion into the host that dissolves cellular tissue and offers them a meal.

An intensely itchy rash develops within three to six hours after exposure to chigger bites (although occasionally the small telltale bumps are not noticeable until a day or two after exposure). After a few days of feeding, chiggers usually drop off the host's body, leaving behind red welts. Avoid scratching the itchy welts, which may break the skin and result in secondary infections.

The best treatment for chiggers is avoidance—which is not always possible. Chiggers are often problems in new home sites while lawns are being cultivated, and they are prevalent in wooded areas, especially during late spring and early summer. If you know you are going into chigger-infested terrain, protect yourself. Insect repellants may be effective (look for ones that contain DEET or pyrethrum). Wear loose-fitting clothing of tightly woven material, making sure to keep collars and cuffs buttoned. Remove and launder your clothing as soon as possible after exposure. A warm soapy shower or bath within an hour or two after exposure will dislodge any chiggers that are attached or waiting to break through the skin. Bathing may reduce the irritation, although once itching has started, it does little more than temporarily soothe irritated skin. The best treatments are antiseptic, anesthetic, anti-inflammatory, and analgesic ointments.

SCABIES MITES

This is another little critter, formally called *Sarcoptes scabiei*, which lives all over the world. It is related to the mites that cause mange (a parasite affliction) in animals. The scabies mite is very small (0.2 to 0.4 millimeters long), and is very host specific: It depends on humans for its life cycle. It is also contagious, and when it is not recognized and controlled right away, it can spread through an entire household. Signs of scabies infection may not appear until a month after infestation. At this point, scabies victims will notice intense itching and slightly raised wavy lines on the skin about half an inch long, sometimes with a pimple at one end. The lines may appear anyplace on the body, but areas with natural creases—waistlines, backs of knees, insides of elbows, for instance—are favored by these creatures. The lines are the burrows created by the female mites. Unlike chiggers, scabies *do* crawl under the skin, where they lay their eggs. These hatch in a few days, emerge from the skin, and molt through several stages into adults. The entire life cycle is carried out on human hosts and lasts about ten to seventeen days.

Although it's pretty easy to diagnose scabies based on the itching and the burrows, it's a good idea to get more evidence. A doctor can take scrapings from the burrows and inspect them under a microscope, for example. This will rule out other forms of skin problems that may look like scabies. The best treatment is with over-the-counter scabicide creams. These usually contain permethrin, lindane, or a similar ingredient. The cream is applied and left on the skin for fourteen to forty-eight hours (be sure to follow the packet instructions). Afterwards, it's a good idea to take a cleansing bath. Although the itching may last for another week or more, this doesn't mean that the treatment is not working. Some patients find that a regular application of corticosteroid cream may help to reduce the itching.

TICKS

Wood ticks—also known as dog ticks—were unpopular in the 1950s, even before Lyme disease was in the common vocabulary. Ticks were thought to be scary creatures. They violated your protective skin to suck the life's blood from your body. They were even worse on your dog, because they concealed themselves in its fur until they were bloated. Sometimes they stayed undetected until they were full of blood and then fell off the dog onto the floor. There they would lie, large liver-colored pearls easily squashed by the bare foot of a child running through the house.

It was not until the saga of Lyme disease began to unfold in the 1980s that we realized that ticks are not just ugly, but dangerous as well. So we soon learned their names, and which ones are serious threats. We found out that wood ticks (*Dermacentor* spp.) and Lyme ticks (*Ixodes* spp.) are different, but we learned too that the latter, also called deer ticks and blacklegged ticks, can transmit the bacteria that cause Lyme disease as they feed on us. Ticks also spread Rocky Mountain spotted fever, babesiosis, Q fever, and tularemia.

Lyme Disease

Lyme disease affects thousands of people each year, and it has been reported in at least forty-eight states. It is actually caused by a spirochete (corkscrew-shaped) bacterium (*Borrelia burgdorferi*) that needs a host in order to grow and survive. Hosts can be deer and mice, and, unfortunately, humans, but typically, those hosts are blacklegged ticks. All ticks are arthropods, related to spiders, mites, and fleas. When they sup on us, it's called a blood meal. Ticks take a long time to eat their fill of blood—sometimes days or weeks—and they feed on more than one type of animal. Some ticks in the larval stage feed on mice or voles and then graduate to larger mammals such as dogs and humans when they are older. (Ticks were apparently put on earth to make other insects look good!)

Spirochetes, including the one that causes syphilis, are known for their ability to migrate deep into the body where our immune responses (and sometimes antibiotics) do not easily reach. This means that they are a serious health threat. Unlike Lyme disease, the syphilis spirochete does not have a nonhuman vector, but is transferred directly from person to person. With Lyme disease, the bacteria needs help to reach its final destination. It is not easy for this bacterium to survive. It needs someplace to live—a host reservoir that has an immune system that does not kill it. Second, it must not kill the host, or it will die, too. And since the host is bound to die sooner or later, the bacteria needs a vector (a carrier) to take it to another host so it does not die off with the first one. After centuries or millennia of trial-and-error living, it now has a well-established life cycle that includes white-footed mice, deer, and blacklegged ticks. (Other woodland hosts include rats, chipmunks, and even some birds.)

Prior to 1975, it was already known that ticks carried *B. burgdorferi*, in addition to the bacteria *Rickettsiae* and even some viruses. In 1975, a group of concerned parents in and around Lyme, Connecticut, spoke with officials at their state health department. They felt that an unusually large number of children had been diagnosed with inflammatory arthritis. Some of the children even had the rare and non-epidemic juvenile rheumatoid arthritis. By the end of 1976, researchers at Yale had found that it was a complicated disease, and it was designated with the new term, Lyme disease. In fact, it appears that bones from individuals two thousand years ago that had shown evidence of rheumatoid arthritis may actually have been victims of Lyme disease.

Lyme disease now accounts for more than 95 percent of all reported vector-borne illness in the United States. That's a total of nearly fifteen thousand cases annually. Most of these cases have occurred in only ten states: New York, Connecticut, Pennsylvania, New Jersey, Wisconsin, Rhode Island, Maryland, Massachusetts, Minnesota, and Delaware. Clearly, the areas of concern are the northeastern and mid-Atlantic seaboard and the mid-northern states, but some counties in northern California are also at risk. The number of reported cases has increased twenty-five times since government surveillance began in 1982. Overall, the incident rate is currently about five per hundred thousand, but it may reach 1 to 3 percent in communities where the disease is endemic. The highest reported rates of Lyme disease are in children two to fifteen years old, and in adults aged thirty to thirty-five. Lyme disease is rarely, if ever, fatal.

The Creepy Crawlers

Fortunately there is some reassuring news. The tick season is short—about three to four months out of every year. And only 1.2 percent of persons with recognized Lyme tick bites actually develop the disease. This is usually because they removed the tick before it had time to transmit the disease (usually within twenty-four hours after attachment). Furthermore, even if they do not get quick antibiotic treatment, 20 percent of victims will only develop the telltale "bull's-eye" rash without other more serious effects. Among those treated, however, about 10 to 20 percent will go on to develop chronic arthritis. Most people who contract Lyme disease and receive quick antibiotic treatment will recover completely, although some of the symptoms may linger for a few months. Even in patients who have developed chronic arthritis, they rarely have continued inflammation of the joints for more than several years.

In general, blacklegged ticks look like other members of that family; they are teardrop-shaped and have eight legs, although the larvae have six legs. Their lumpy heads are not really heads: they have rudimentary eyes and sophisticated mouthparts in order to do their sucking. They are, however, sensitive to warmth and have a keen sense of chemicals. In fact, they are attracted to carbon dioxide. That's how they detect humans and other mammals coming along a path in the woods.

Blacklegged ticks also differ from wood ticks (also called dog ticks) in several ways, but mostly in size. Wood ticks are about three to four millimeters wide and a bit longer. An adult blacklegged tick is half that size, or about the size of a sesame seed—about two millimeters. A blacklegged nymph is the size of a poppy seed, about one millimeter, less than half the size of an adult.

Blacklegged ticks and wood ticks also have different color markings, which can best be detected with a magnifying glass. The blacklegged tick is brown or black with eight black legs, and has an orange-red crescent on its back (dorsal shield). An adult wood tick is brown with white markings on its back, and its underside is tan with brown spots.

There are two recognized species of this creature in North America: the black-legged tick and the western blacklegged tick. To the human eye, however, they look the same. What is important is that both species are carriers of Lyme disease, and both bite humans. Remember that this disease is not transmitted from person to person, nor is it transmitted by flies, fleas, or lice. While blacklegged ticks can and do crawl, they do not jump onto potential hosts, nor do they fall onto them. They contact their hosts by "questing" for them. This is a scientific term for lying-in-wait. At all three stages—larva, nymph, and adult—the blacklegged ticks set themselves up on a blade of grass or other prop, hang on their back legs, and hold out their front legs. When an appropriate host brushes by, one that exudes warmth and carbon dioxide (which the tick senses with an organ in its front legs), the tick lets go of its perch and latches on for a ride with its new host.

A tick tends to climb against gravity. It might attach at the site of contact, or move about until it meets an obstacle. It usually looks for a spot on its host where it is least likely to be detected, such as in the hair or in folds of the skin. Once it has found a good location, it sets about obtaining its meal. First it secretes a sort of cement that adheres it to the host. That is why an attached tick is difficult to remove. It then uses its mouthparts to create a little well in the skin of the host. Blood collects there, and the tick drinks it. When the tick has had a full meal, it falls off.

Blacklegged ticks are actually quite fussy about where they live. Here are some of the conditions they prefer:

- High humidity environments, such as rivers and coastal areas.
- Shaded, damp, forested, and brushy areas covered with leaf litter.
- Areas frequented by deer: woodlots, wooded areas between yards, edges between yards, and wooded or unmaintained brushy areas.
- Areas frequented by mice and chipmunks, such as woodpiles, stone walls, unkempt bushy areas, and tall grass at the edges of residential properties.
- Brushy and grassy areas along hiking trails in areas frequented by deer.
- Bayberry, scrub oak, and other scrub habitat in northeastern coastal fringe areas.
- Oak woodland and chaparral ground-cover habitats of coastal northern California.

Here are some suggestions on how to prevent Lyme disease:

1. It's important to know your habitat and to be especially cautious in wooded areas in the regions where Lyme disease is endemic. Be aware of the seasons when the risk is highest. Blacklegged tick nymphs usually take their blood meal from late May through early July; western blacklegged tick nymphs take this meal from March through June.
2. Wear good protective clothing in the woods. Light-colored long pants are best, and these should be tucked into white socks. This keeps ticks away from the skin, and the light color helps you to spot them if they cling onto you. Dusting your clothes with botanical powders containing pyrethrum will help, too, or try repellents with DEET, citrus oils, or insecticidal soap. Try to stay on well-beaten paths or in clearings. At the end of a hike, remove your clothes and wash them if you are in a high-danger area.
3. It's also helpful to use one of the commercial repellents. Some of the popular brands include HourGuard12, Off! Deep Woods, and Repel Permanone.

Check the *Consumer Reports* analysis of these products for the one that will work best in your area. As for pesticides, these are not recommended; they are highly toxic and may leach into your ground water supply. A better solution is to remove any woodpiles where mice may flourish, and to discourage or prevent deer from entering your yard. Keep the grass mowed and clear any shaded areas, opening them to light. Remember, blacklegged ticks like moist, shady areas.

4. Check for ticks regularly (especially on children), and remove any offenders quickly and carefully. Be sure to use tweezers and make sure you don't accidentally inject tick fluids into your body. Use a pair of tweezers to grasp the tick as close to the mouthparts as possible. Pull straight up slowly and steadily until the tick is removed; then treat the bite with alcohol or iodine. Destroy the tick by drowning it in iodine or soapy water, but remember that you'll expose yourself to any tick-borne diseases if you squish it between your fingers. Don't leave a dead tick attached and do not cover the tick with toothpaste, Vaseline, or anything else waiting for the bug to smother; you'll only prolong any exposure to the disease. If you're willing to risk a burn, you can also use a hot needle to kill the tick.

Don't forget to check your pets for ticks in the fall and winter, and don't neglect their ears and feet, where ticks like to hide. Ticks and nymphs (the immature tick) are most active in the spring and summer, although some adults are active through the winter. The brown dog tick (*Rhipicephalus sanguineus*), a dog's main enemy, will move into your home and lurk about in cracks and crevices waiting for a ride and a meal, probably with the dog. Ticks that carry Lyme disease are obviously of the greatest concern.

Once contracted, Lyme disease has no cure (for humans or for animals), but the disease can be arrested with antibiotics. Because Lyme disease transmission has a time factor, the sooner you remove the parasite, the lower the chances are of getting sick.

Other Tick-Borne Diseases

Ticks are ideal hosts for all kinds of germs, including bacteria, viruses, and parasites. The most distressing effect is clearly Lyme disease, carried by the hard-shell ticks. But some of the soft-shell ticks are a health hazard, too. Prior to the identification of Lyme disease, Rocky Mountain spotted fever was probably the most commonly recognized tick-borne illness in North America. It gained notoriety in the second half of the nineteenth century in Montana, when settlers were tormented by the disease, which is sometimes fatal. It's caused by a bacterium called *Rickettsia rickettsii*, discovered by Dr. Howard Taylor Ricketts in the early 1900s.

In spite of its name, Rocky Mountain spotted fever is primarily a disease found in the Southeast and in the west south-central United States. It usually produces multiple symptoms, such as fever, headache, and rash, but other effects might include abdominal pain, mental confusion, and organ failure or dysfunction. The disease is especially common on the Atlantic seacoast, and occurs most often from May to September, when people and ticks are active outdoors. Fortunately, prompt treatment has led to a reduction in the number of deaths from 25 percent of cases to only 5 percent. Nonetheless, it may take months for a victim to recover fully.

Blacklegged ticks and western blacklegged ticks also carry the recently detected disease, babesiosis. It's caused by a parasite that infects the red blood cells. Although it is occasionally confused with Lyme disease, it is more closely related to malaria (another parasitic infection). The symptoms are sometimes hard to detect, but they may include fever and anemia. It most commonly occurs in the Northeast during the summer months. An antiparasitic drug is used to treat babesiosis, but even without treatment the disease usually goes away by itself. It is rarely fatal.

2

THE FLYING INSECTS

FLIES

Most people have heard the famous folksong "Shoe Fly, Don't Bother Me," and we all know exactly what it means. These little insects can be incredibly annoying, both inside the house and in the great outdoors. Flies are one of the largest groups of insects in the animal world. In fact, there are about 85,000 kinds of flies in existence, and more than 16,000 varieties live in North America. No wonder they are such nuisances.

Flies, members of the order Diptera, are small insects with a single pair of wings, relatively soft bodies, and mouthparts that are designed for sucking. Because there is a vast array of species, it's easier to deal with them if we divide flies into their two sub-orders, the Nematocera and the Brachycera. The long-horned flies (Nematocera) are some of the smallest and most bothersome of this family. They include blackflies, mosquitoes, midges, gnats, and no-see-ums. The short-horned flies (Brachycera) are the big and aggressive ones, such as the horseflies and the deerflies. The sub-order Brachycera also includes the familiar houseflies and the fruit flies.

Flies eat a variety of meals, including nectar, fruits, vegetables, and meat left in the open, and—in the case of females—human blood. You will find them across most of North America, although they tend to thrive in warmer climates, or during the warmer weather of more northern regions. Mosquitoes are almost everywhere, but they are particularly fond of habitats that are near water. That's where the larvae grow before they move into their next stage. Horseflies and deerflies like to live in fields or forests, especially if they're near lakes or rivers.

And you'll certainly find them around farms or anywhere there are cattle or horses. Houseflies, of course, are everywhere, too. Their friends, the fruit flies, are magnetically attracted to any rotting fruit.

The bad news is that flies often reproduce quickly and go through some of their growing stages at a remarkable pace. Fruit flies move through their four life stages (egg, larva, pupa, and adult) at such an alarming rate that they are frequently used in genetic tests so that scientists can get quick results. The larvae thrive in water, but they are just as happy in decaying wood or plant matter, and even in other animals. You'll know these delightful little creatures by their more common name—maggots. If you've ever seen a dead animal on the road or in the woods, then you've probably seen the second stage of these creatures. It's pretty gross.

Flies do perform some important functions in the natural world. A few of them prey on other insects, so they help to keep populations stable; however, many of them bite us and our pets, making themselves real nuisances. And a few of them damage our plants, carry diseases to us, and some even carry plant diseases around, too.

While they may look fairly harmless, flies harbor disease. They breed in areas that are particularly filthy, so it's not surprising that they've been known to carry such diseases as typhoid fever, dysentery, and cholera—to name only a few. That's why it's distressing to see how common they are in some restaurants. If you see a lot of flies in a restaurant, it's a good sign that the establishment won't get four stars when the health department comes around. On the other hand, if you see absolutely no flies in a restaurant, it's a good guess that the place uses a lot of pesticide. Take your pick.

Horseflies and deerflies are the giants of this insect order. The females are the bloodsuckers, while the males (rarely seen) are off feeding on flowers. These are the critters that you'll see around horses or cattle, driving them crazy. They are also the ones that seem to like to attack you just when you've come out of the lake or pool after a swim. The reason is that they breed near water, and that's where the larvae grow. The horsefly is the larger insect, while the deerfly is closer in size to your average housefly. All of them are really strong fliers—as you'll know if you've ever tried to chase or whack one.

The most obvious way to avoid the housefly problem is not to leave any food or water sitting unattended. Yet, let's face it, common houseflies will get in your house from time to time, and they're annoying. Sometimes, if you're fast, you can do them in with a fly swatter or a rolled-up magazine. Flypaper is unattractive, but it works and doesn't harm people living in the house, as would an indoor spray. Today you can even buy sticky flypaper that's hidden inside a decorative box so you don't have to look at the dead flies all day.

The Flying Insects

Be careful if you try more "aggressive" methods. When you use a pesticide-infused No-Pest-Strip hung in the middle of a room, it certainly kills flies, but its presence means that you're inhaling pesticide vapors all day and all night long. Avoid aerosol pesticides because they may get into your lungs, too.

New England residents have to deal with cluster flies, which hibernate in the walls of houses and emerge inside warm rooms on sunny days. One resident tried a benign method: a Venus flytrap on the kitchen windowsill. Nevertheless, it won't make much of an impact alone on the fly population. Other New Englanders merely vacuum the pests up during the colder months of the year and toss them into the frigid winter air where they die.

Fruit flies seem to appear out of nowhere as soon as anything in your fruit bowl begins to ripen—especially if the room gets too warm. This small (quarter-inch long), often striped, and brightly colored species is very common and it thrives on flowers and any kind of vegetation. You don't even need to have fruit in the house to attract these little gnats. They can reproduce on anything that is wet. They might be breeding on the damp pipes under your kitchen sink, or in puddles of moisture in the basement. After they emerge as fully grown flies, they

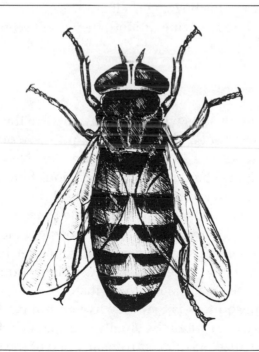

Short-horned fly (Brachycera)

head for areas that are damp and well lit. That's why the kitchen—and the fruit bowl—is a favorite target.

Here's a tip from the entomologists at Iowa State University. If you're worried that they're breeding in your drains, cover it with a bit of plastic film taped to the hole. In a day or two, check it to see if there are flies attached to the surface. That will give you a good clue about their numbers. You may simply need to give these drains a good flushing or cleaning with a heavy-duty cleanser.

The Mediterranean fruit fly is the most serious pest when it comes to citrus fruit. Other varieties prefer bananas and peaches. The obvious solution is to keep your fruit and vegetables refrigerated, or at least in a sealed and cool part of the kitchen. Sprays don't usually have much of an effect on these flies, and they tend to disappear of their own accord within a few days.

In the great outdoors, traps can help to reduce the number of flies in the area. Peaceful Gardens, an organic garden supplier in California, sells a flytrap that attracts insects with smell—a strong ammonia odor. This trap is designed to be used around animals, say, in a horse barn, and it's so smelly that you can't bring them in the front door of a house. But they work well if they're far enough away from your living quarters.

The flies are attracted to the ammonia smell of the bait in the trap, fly up to it, and can't get out. Says Mark Fenton of Peaceful Gardens, "After you get a couple thousand flies in the trap, the buzzing itself attracts others from a long distance." Even with all the dead bodies lying around, they don't seem to get the message.

MOSQUITOES

These small, soft-bodied insects with their long legs are the bane of outdoor living. There are actually about 3,000 species around the world, and about 5 percent—or 150—of those species live in North America. New Jersey is frequently called the "Mosquito State" because it has the dubious honor of hosting about 63 species. Females of the species *Anopheles* have the ability to transmit the malaria parasite, and this is one of the species found in the state. New Jerseyans even divide them up into the major breeding habitats: snowpool mosquitoes, floodwater mosquitoes, swamp breeding mosquitoes, and container breeding mosquitoes. There's an old saying in the state that it's possible mosquitoes drove the first Swedish settlers from the state and not the British.

The myth about mosquitoes is that they live on human blood, but that's not true. Their main source of nutrition is actually sugar, which they get from the nectar of plants, from the juice of fruits, or from any one of the sweet secretions that may seep out of plants. Just like athletes, this energy hit gives them the strength to do all of their flying (they may travel as much as 150 miles in their short, three-week lifespan). Mosquitoes need lots of sugar every day, but the females also need some blood for their egg-laying work.

Mosquitoes need blood in order to procreate. The blood actually helps the female mosquito to produce fertile eggs. (Because the males don't lay eggs, they don't need blood and therefore don't bite.) The females head off every day on what's called a "host-seek" in order to get a blood meal. You won't be surprised to learn that they can lay as many as three hundred eggs at a time, each brood of eggs requiring a human bite. And these eggs can last in a dried state for as long as five years. No wonder it feels as if they're everywhere.

So it's the females that are the real culprits. They have mouthparts that look like little stiletto heels, which make them serious pests for you and your pets. They not only bite us and drink our blood, mosquitoes carry various diseases, including the aforementioned malaria, as well as yellow fever, dengue, filariasis, and encephalitis. Many of these medical problems are found only in the tropics, but they have been known to migrate to North America.

When a mosquito bites you, it secretes a bit of saliva into your blood to keep it from coagulating. It's this saliva that makes the bitten area swell up, not the bite itself. Then, when you scratch it, the bacteria from your fingers or under your fingernails often get in the skin. That's when the area becomes infected and you see the swelling. Why some people are more attractive to mosquitoes than others is still a mystery. In a few cases, it may have something to do with their perfume, their choice of soap, whether they perspire heavily, or even if they have nervous personalities.

Female mosquitoes are attracted to us by the carbon dioxide we give off. They can detect a little patch of unprotected skin as small as a dime—and they can do that from eighteen feet away. When they land on you, they then have to decide if you're a tasty meal or not. Apparently the folic acid on our bodies is one of the clues they look for, but this smell could be balanced by the other odors we've just mentioned, or even by the smell given off by the detergent used to wash our clothes. It's a complicated decision that the mosquito makes, although some people certainly feel that they are high on the "mosquito meter."

Mosquitoes need water; that's where the larvae live. So you'll find them breeding in rivers, lakes, and marshes, as well as in your swimming pool or even in a tree if it has an area for water to collect. The eggs are laid on the surface of the water, the larvae then live on organic matter, and the pupae race around in the water before they become full adult mosquitoes. The obvious way to combat them is not to leave any standing pools where they can breed.

Many species only live for a few weeks, but a few can last as long as two to three months. It's encouraging to know that so many other creatures like to eat these critters. In the water, mosquitoes are a favorite food for many kinds of fish. Once they're in the air, they are devoured by a variety of birds—including swallows, martins, and bats. You'll see these creatures darting through the summer air as they consume our little enemies. Building nests for these birds is, in fact, one way to begin to control mosquitoes on your property. As for their living conditions, mosquitoes thrive in eighty-degree weather, because they are cold-blooded creatures and they maintain the same body temperature as their surroundings. That's why their numbers really do increase or decrease with the changes in weather.

Mosquitoes can transmit diseases from one human to another. Malaria is one potential problem that's spread by these insects. Encephalitis can be transmitted from infected birds to us, and yellow fever can migrate from monkeys to humans via mosquitoes. (Tests have shown, however, that the AIDS virus cannot be transmitted from one human to another through mosquitoes.) Most of these medical problems are covered by the common vaccines we give to children and adults.

Mosquitoes are certainly one of the most pesky critters around, but many deterrents are distasteful. Some are foul smelling and contain DEET, a chemical that may yet prove to be carcinogenic (DEET is in some of the more popular outdoor repellents). Then there are the bug zappers, which are indiscriminate in their effect, and kill all insects, good or bad. When you're sitting in the backyard listening to the rapid *zap!* of the zapper, you're hearing good insects go up in smoke, too. The zapper's light also attracts all the bugs in the neighborhood to your yard. If you want to use a zapper, give one as a present to a neighbor a few houses away. Backyard foggers and sprays also kill indiscriminately, and expose you to pesticides.

What are we to do? There are several ways to keep mosquitoes at bay. You could stay indoors or in a screened porch. But you can't always keep away from mosquitoes. Besides, from time to time one mosquito will get under the covers with you, and head straight for your inner ear. Insect repellents are probably the most widely used weapon against mosquitoes. Some that are based on natural ingredients have recently been introduced to the market. Cal Saulnier of Plow and Hearth Catalog says, "The natural insect repellent, made from pennyroyal and citronella, tested pretty well, and with the controversy surrounding DEET, it can't hurt to try it."

What's wrong with DEET? It's hard to know. DEET is in most of the repellents you buy at the drugstore. People have been using it for about thirty years, and most of us don't think twice before we spray on the Off! repellent. However, the Environmental Protection Agency has issued warnings about possible side effects. These may include headaches, mood changes, confusion, nausea, muscle spasms, convulsions, or unconsciousness. Children, it seems, are particularly vulnerable. The agency has spent considerable time looking at other more serious side effects. They've investigated birth defects, cancer, and reproductive problems. It's important to know what you're using when you apply this stuff.

Still, the most effective repellents around still contain DEET, although you don't need to use anywhere near a 100 percent solution for it to work well. A few years ago, *Harrowsmith Country Life* magazine carried an article on natural repellents. They found that Bug-Off, Cedar-al, Natrapel, and Avon's Skin-So-Soft all were effective, although none of them worked as well as repellents that contained DEET. The natural varieties use volatile oils, and the effectiveness simply wears off quickly.

Remember that any repellent works best if sprayed on, not rolled or rubbed on, and be sure to treat your clothing as well. (Do remember that DEET will melt plastic, so be careful where you spray it.) Be sure never to use any product with more than 14 percent DEET on children. And never use DEET products on newborns.

Here are some low-tech solutions you might want to try, too. Citronella-scented candles or pyrethroid torches and candles all work well as mosquito deterrents. Netting is also a smart thing to pack if you are headed to the north woods in late spring or early summer, where these pests will be especially numerous and aggressive. You can buy mosquito-netting hats from several outdoor outfitters, but be sure to treat the hat with repellent. Wood lore has it that eating large quantities of garlic can keep mosquitoes away. Some people even use naphthalene granules, a product purported to repel everything from deer to squirrels, and of course moths, since naphthalene is the main ingredient in mothballs.

On the higher-tech end of the spectrum are ultrasonic insect repelling devices. The plug-in kind is supposed to mimic the sounds of the dragonfly, a natural enemy of mosquitoes. Other devices are supposed to sound like male mosquitoes seeking a mate, and the females are therefore sometimes repelled. But some entomologists are skeptical, as are some outdoor products industry professionals.

Wayne J. Crans, an associate research professor at Rutgers University, says, "Electronic mosquito repellers do little in the way of reducing mosquito annoyance." He's found that "female mosquitoes do not flee from amorous males, and mosquitoes do not vacate an area hunted by dragonflies." He points out that the merits of these products are rarely backed with any kind of scientific testing or evidence.

Sonic Technology's Lowell Robertson agrees that tests have proved disappointing. "Early on, we did extensive testing on mosquitoes, because mosquitoes are a high-visibility insect and one that people would definitely like to get rid of," he says. "We found out ultrasound isn't a very good tool for getting rid of mosquitoes." And Cal Saulnier says his company, Plow and Hearth, stopped carrying an ultrasonic insect repeller, discouraged by the ten percent return from dissatisfied buyers.

Professor Crans is also not a fan of bug zappers. These are outdoor electronic devices that lure bugs in and then electrocute them. If you've seen them in action, you'll be familiar with the zapping and sizzling sound they make as the little insects get fried. The problem is that it appears they're doing more bad than good. "Bug zappers kill a lot of insects," says Crans, "but very few of the insects killed function as pests." In fact, most of the "popping" sounds are actually moths flying about at night, when they're naturally attracted to the light. The longer, drawn-out "sizzles" are usually beetles.

The statistics that the staff at Rutgers has collected is pretty convincing. "Comparison trapping," as they claim, "has also shown no significant difference in mosquito populations in yards with and without the traps." The number of biting insects killed by these electronic zappers is actually less than 1 percent of anything killed each night. It seems that the owners just like to hear the sound of things being killed, because it appears that they're doing something about the problem. In reality, they may be making it worse.

Here's another "solution" that seems to be suspect. There's a relatively new plant on the market called the Citrosa. It's been genetically engineered by crossing the grass that produces citronella oil with a variety of geranium. The resulting plant is supposed to produce a smell that will get rid of mosquitoes. Tests have shown that the Citrosa does little or nothing to repel our little pests. The mosquitoes even landed right on the leaves of the plant and weren't discouraged at all by the smell!

Dr. Arthur Tucker at Delaware State College has found that this "miracle plant" contains only 0.09 percent citronella. There are better choices if you want to grow plants that will discourage mosquitoes. Lemon thyme herb is one good alternative. But the fact is that these plants only release a small amount of the fragrance that discourages mosquitoes. A better solution is to take the leaves off one of these plants, crush them, and rub the mixture on your skin. Just make sure you check for an allergic reaction before you adapt it as your new favorite repellent.

Another way to cut down on mosquitoes is to install a bat house in your yard. A single bat can consume up to six hundred insects an hour. Nevertheless, the folks at Rutgers have once again asked us to temper our enthusiasm for this "solution." Bats are voracious insect eaters, but they will eat anything that's available. So if you have lots of mosquitoes in your yard, they'll eat a large proportion of them. But they will generally behave pretty much like the electronic zapper; they'll kill anything that's available.

It may also be helpful to replace your outdoor lights with the newer "bug" lights that are yellow. They tend to attract fewer night insects to your home or yard. You may have to special order these devices if they're not available at your local hardware or lighting supplier.

MOSQUITO BREEDING GROUNDS

Tires are their preferred home base, but Rutgers staff has also isolated these breeding sites:

- *Clogged roof gutters.* These need to be cleaned on an annual basis, because they provide another moist, bacteria-filled environment for breeding.
- *Stagnant water.* Turn over unused wading pools and wheelbarrows so they don't become convenient places for the eggs to multiply.
- *Ornamental garden pools.* If you own one of these otherwise attractive features, make sure that it's aerated or it has hungry fish in it. Chlorine also helps.
- *Swimming pools.* Cover your pool if you're going to be on holiday for an extended period, otherwise you're just breeding mosquitoes for your neighbors. Evidently a mosquito will easily identify any water that's been left standing for more than four days—and will move right in.

Purple martins (*Progne subis*) are usually thought to be the "natural" solution for mosquitoes, but that's not entirely true either. Although they eat lots of the little devils, purple martins also consume dragonflies, which are the most voracious eater of mosquitoes. Martins do eat mosquitoes, but not nearly in the quantities that the mythology suggests.

Speaking of dragonflies, these often colorful insects are great consumers of mosquitoes in their adult (biting) stage. Together with their relatives, the damselflies, these large four-winged speed demons have a unique method of capturing their prey. They make a basket-like trap of their legs as they fly, catching the mosquitoes in midair. Dragonflies are found across most of the middle and northern parts of the United States and into the southern parts of Canada. The damselfly tends to favor the eastern half of the United States.

Still another tactic is to attack mosquitoes at the source: Cut down their populations by getting rid of their breeding grounds. Mosquitoes aren't picky about nurseries. All they need is a small amount of water: a flower pot, a fish bowl, a garbage can, a birdbath that hasn't been changed for several days, a pool of water behind your garage. Even the smallest puddle of water can give mosquitoes a place to breed.

TOTAL CRITTER CONTROL

The best time to hunt for stagnant water is after a rainstorm or after you water the lawn. Trash can lids and wheelbarrows are places where water can collect. Empty them immediately. Then try to set some traps. Leave a pool of water around, perhaps in a soup bowl or flowerpot—but add soap to the water. Female mosquitoes will land on the water to lay their eggs, but these mothers-to-be will die in the process when they drown in the soapy water. If you have a fishpond, you can even stock it with mosquito larvae-eating fish. Trout and bass are good bug-eaters, for example. The bacteria strain *Bacillus thuringiensis* (Bt), used by organic gardeners against pests, will also kill mosquito larvae, which eat the bacteria and die.

Researchers at Rutgers have found that discarded tires have actually become the single most important habitat for breeding mosquitoes in all of North America. These old hunks of rubber tend to create little pools of water. Because they are black, they also tend to stay warm. Add in that they're a form of shelter, and you've got a virtual manufacturing plant for mosquitoes. Ridding your property (or your neighbor's) of all used tires is the first step in bug prevention.

Finally, mosquitoes are generally night feeders, although they prefer to eat at dawn and dusk. Their activity increases an amazing 500 percent on nights when there's a full moon. You might want to plan your camping trips—and your dashes to the outhouse—accordingly.

BLACKFLIES

Also known as buffalo gnats or turkey gnats, these little critters are incredibly annoying to anyone who enjoys the great outdoors. Although many varieties only live about two or three weeks, they breed so prolifically it seems as if there are always millions swarming around your body. Blackflies grow in running water, which means they tend to like the same rivers and streams (and nearby lakes) that we enjoy for our vacations. They live on the bottom of the fast-running water, and then rise to the surface in a little bubble of gas to emerge into the air as a fly.

These hump-backed flies, about one-quarter of an inch long, are similar to mosquitoes because only the females like to drink blood. The males live on the nectar of plants. They are unlike mosquitoes because they feed in the daylight, while their bigger cousins prefer to eat in the evening or at night. It's almost as if they organized their shifts in order to torment us. Blackflies prefer to bite you on the head, but they also enjoy little nibbles on the ankles and wrists.

When the female bites you, she injects a small amount of anticoagulant through her saliva. This keeps your blood from clotting, as it would if you cut yourself. For the blackfly, this injection allows the blood to keep running into her mouth; however, this same saliva is what causes us such irritation. It will make your skin swell, particularly if the bite occurs around sensitive areas such as your

BATTLING BLACKFLIES

- Wear light-colored clothing when you're outside. Blackflies are attracted to dark colors, so try to wear white or tan hues to make yourself less of a target.
- Always wear thick socks, and tuck your pants into them to prevent the blackflies from getting at your lower legs.
- Use insect repellents on all exposed areas of skin. The neck, for example, seems to be a favorite target. Some hikers have found Muskol to be one of the many effective deterrents. There are a number of similar products available for pets and livestock.
- When you are camping or hiking, try to pick your sites away from the obvious blackfly breeding areas we've described.
- If you have to work outdoors during the critical seasons for blackflies, you might want to invest in some specialized clothing or netting to keep these pests at bay.

eyes, ears, mouth, or nose. The swelling may even last for a few weeks, and the affected area will be very itchy. Some people are actually allergic to these bites. There have even been a few rare cases where blackflies have caused death from their venom, as a result of toxic shock to the body.

It would be almost impossible to control all of the potential breeding sites for blackflies, particularly if we don't want to harm the environment. Blackflies also tend to migrate far from their breeding areas, so it's hard to know where they even started in most cases. The good news is that blackflies evidently do not carry diseases to humans. In that sense, they are more of a pest than a threat. But they are certainly a nuisance to us when we're outdoors, and they are equally annoying to cattle and horses, and to our pets.

NO-SEE-UMS

The nasty little biter we call a no-see-um is also known in some areas as a biting midge, a gnat, a punky, or a sand flea or sand fly. Under a magnifying glass, it looks like a small mosquito, except that its stinger is a bit shorter. The body parts are brown, the two narrow wings are translucent but sometimes hairy or spotted, and it appears whitish brown when it's flying. They also seem to live everywhere on the planet.

Life for a no-see-um begins as an egg laid by the female in the spring, often in marshy pools where there is rotting vegetation. Any little bit of water will suffice, however, if there's no ideal habitat. (They like saltwater and fresh water equally.) Sometimes the young are even buried in up to four inches of mud in intertidal areas. The eggs then progress through the larval stage into the pupa, and these float like little corks on the water. We can be grateful that frogs, crabs, fish, birds, and other insects all like to eat these little morsels. If they survive, a week later they turn into sand fleas and start buzzing around your head. Scientists believe that about 90 percent of all the no-see-ums that arrive at the adult stage are eaten by predators, mostly by birds.

Although the no-see-um is only one-tenth of an inch long, the bite of this tiny vampire can cause real discomfort for the victim. It's actually more of a sting than a bite, and the effects can last for hours. You'll see a little welt forming on your skin about the size of a freckle, and it may be red or white. Usually there's not just one, but many, because the no-see-ums travel in swarms.

Some varieties seem to be even more vicious than others. In a few of the equatorial countries where North Americans like to vacation, you may find these nastier

---

CONTROLLING NO-SEE-UMS

- These bugs are at their busiest on beaches in the evening, or when the wind dies down. If you happen to be around at those times, you'll need to take further precautions.
- The most popular types of repellent use the DEET formula.
- Other skin applicants that seem to get good results are cactus juice, or the Avon product Skin-So-Soft.
- Long-sleeved shirts and pants are obviously a good idea for walks on the beach in the evening if you want to keep your exposed skin to a minimum.
- You can also spray your clothing with some of these repellents, or try one of the bug-gear clothing lines that outdoors stores promote.

---

species. Scuba divers have reported beaches where they were literally driven indoors by the swarms. There have been accounts in diving journals of people who've suffered from severe fever, rashes, body parts covered in red welts, facial blemishes, and even cystic acne while holidaying in Honduras. It's possible that some people will have a severe allergic reaction, and could even die from acute kidney failure.

A few of the other destinations where you'll have to be on the watch for these strains of no-see-ums are Belize, Trinidad, Hispaniola, and further afield in Thailand and Egypt. There were also reports of no-see-ums problems from U.S. soldiers returning from the Gulf War after Operation Desert Storm.

Back at home, these little pests often plague horses. Because of the size of the animals, it's often hard to spot when and where horses have been bitten. They seem to be prone to allergic reactions, which can produce sores and areas where the animal's hair will fall out. No-see-ums can also transmit African horse sickness as well. Horses may need to be sprayed with insecticides, screened by areas coated with repellent, or kept in their stables during the evening hours if they are near infested areas.

BEES, WASPS, AND YELLOW JACKETS

Say "bee" and most people think of the cute yellow insect on a flower. Or maybe they remember being stung when they walked barefoot through clover. Or the great killer bee skits of the classic *Saturday Night Live* television shows. But the truth is that, just like ants, bees and wasps have a well-organized social structure. They are the corporate members of the insect world, with a highly developed system of work and reward. These creatures live in a communal world, where everyone has a specific task to perform. There are more than 100,000 species in the world, and more than 16,000 of them live in North America. And everything works fine, as long as we don't disturb them. Then all hell breaks loose.

Meanwhile, say the word "bee" to a resident of Scottsdale, Arizona, and you'll probably be hit with a barrage of bee stories. It seems that a warm, wet spring in 1991 led to a bee population explosion in the surrounding desert. Then the water dried up, and the bees headed for some of the swankiest pools in Scottsdale. One resident reported seeing as many as ten thousand bees at his pool a day! They clogged skimmers and terrified the pool owners.

There was really no solution to the problem this time, and without the pools, the bees probably wouldn't have made it through the summer. People tried to lure the bees into kiddie pools. No luck. Now every year the residents of Scottsdale hope for a cold, dry spring. But aside from this kind of occasional freakish occurrence in nature, bees are seldom a problem.

There are some pretty obvious differences in the members of this insect family. Bumblebees (family Apidae) are the giants of the group. They're the big, robust, hairy creatures that we see in the garden or woods. They can range from a half-inch to more than a full inch long. Most are yellow and black, although some sport an almost rust color. Bumblebees are found virtually everywhere, including the Arctic, but they are mostly non-aggressive when it comes to humans.

They also tend not to invade our houses or surrounding buildings. The other good news is that they build a new nest each year, so they don't tend to become annual pests—like some of their relatives—once they've set up home. Bumblebees will occasionally take up residence in a wall cavity, or even in a dryer vent in the wall, but mostly they tend to build their nests in the ground or in a tree. Sometimes a woodpile also becomes a site to set up the colony.

Honeybees (*Apis mellifera*) are their close cousins. These bees tend to be a bit smaller and more slender than their larger cousins, but they are also somewhat hairy. Their coloring is a bit less yellow, and there are often more strips of color, as opposed to the solid blocks on the bumblebee. If you get close enough, you will see the little pollen sacs on their hind legs as they go about "accidentally" pollinating the plants and fruits in your yard. Bees, unlike flies, have two pairs of wings—usually with a large set slightly forward of the second pair. They never nest in the ground; their nesting material is actually wax, and this is where the honey is stored.

Wasps, hornets, and yellow jackets are another story. The most widespread of the group are the bald-faced hornet (*Dolichovespula maculata*) and the yellow jacket (*Vespula* spp.), although a relative—the yellow mud dauber (*Sceliphron caementarium*)—is also common across North America. This family tends to have shinier and less hairy bodies. The coloring is similar, with black and yellow being the predominant range, but the wings tend to be longer and more swept backwards than most bees. They are generally folded parallel to the body when these insects are at rest. The other feature worth noting is the actual body itself. Where bees' bodies do not appear to be very segmented, you'll see in the wasp family that the middle and back sections are joined by what looks like a thin wisp of thread.

The larvae of this group can look like caterpillars or maggots, and many of them are plant eaters. They'll live in—and eat—the leaves, stems, or even the fruit of your garden plants. Some of the larvae are parasites, so they'll take up residence in the body of some other creature, including another insect. Others live in nests built by the adults, who then bring them food. Look for paper nests if you're trying to find the adults of this group. They will also nest in the ground. Yellow jackets and hornets create these nests by stripping fibers off exposed wood. That's why you see them hanging around cottages, boat docks, around

The Flying Insects

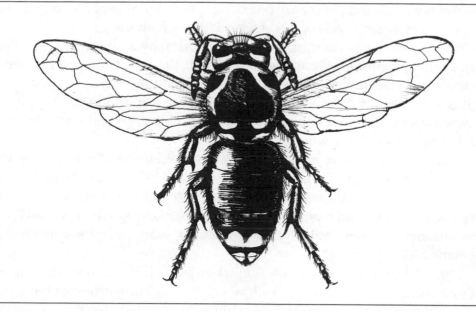

Bald-faced hornet (*Dolichovespula maculata*)

barns, or under the soffits of your house. Even wood fences can provide building materials for these creatures.

It's important to know that some members of this family are really useful to humans. Bees, for example, produce honey and wax for us, and they are the great pollinators of the natural world. Without them, plants and fruit trees would never grow and reproduce. Many members of this family are also insect eaters; they help to keep the population level even. The bad news is that they can, and do, create a nuisance when they eat our plants.

The big problem for us, of course, is that these things can really sting. Only some of the females have this ability. They sport a sharp, hollow organ that is sometimes barbed, and it's connected to a poison gland in their abdomen. If you annoy these creatures, they will jab you with their stinger. It in turn injects you with a poison that will make the wound itchy and painful. Usually, this poison is a kind of formic acid, and the results can be very unpleasant—sometimes even lethal. And more bad news: These guys can sting repeatedly and live to sting another day.

Sure, there are the big bumblebees in the honeysuckle, but they're pretty slow. Yellow jackets and their friends are the thugs of the wasp world. They're the ones that swarm around your head and try to get into your soft drink. They gather around garbage cans, and annoy you at a backyard barbecue. They love sugary foods and proteins, and people are often stung while dining al fresco or accidentally when they ingest a yellow jacket with a bite of hamburger.

Yellow jackets live in colonies started each spring by a lone queen. She begins the nest and lays some eggs, and then after about a month, some workers emerge from the larval state to help expand the nest and feed the larvae. Soon the queen only has to lay eggs and the workers take care of the rest. At the height of the season, new queens and males are produced; they will mate and only the queens will hibernate through the winter.

As it gets cooler, activity increases in the nest, and the workers concentrate on quick energy, high-sugar foods instead of the proteins they sought earlier in the summer. They're more aggressive late in the season. The yellow jackets usually live just one summer, and then they die off, leaving the queen to hibernate. A cold winter kills many queens, but after a warm winter, more queens survive and produce many more offspring.

In general, bees and wasps only sting when provoked. But how do you know what provokes them? For starters, yellow jackets and hornets nest in the ground, in old stumps and logs. If you dig up a nest in your garden, that's certainly a case of provocation. You may be the next victim of a mass attack. These insects may also make nests of a papery material in a protected space, such as a garage, or under a backyard deck. If you accidentally hit it with something, the attacking hordes will be on you. Yellow jackets are the most unpredictable of the bees and wasps and sometimes sting for no apparent reason.

When you're stung by a wasp or bee, wash the wound and then reduce the pain by applying ice, meat tenderizer, wet tobacco, or a commercial sting treatment to the area. If it's a bee sting, first scrape the sting with a plastic credit card, your fingernail, or a driver's license to take the stinger out (bees, unlike wasps, leave a venom sac in the wound with the stinger). Signs of an allergic reaction include: difficulty in breathing, severe swelling, or hives. If you notice any of these symptoms, seek medical attention immediately. Anyone stung in the mouth or throat should get to a doctor right away. Swelling caused by a sting in the neck could easily block your breathing passage. Approximately twenty people in North America die each year from anaphylactic shock after being stung by these pests.

If you keep calm while a yellow jacket is around your head, you may avoid a sting. Don't swat wildly at them, but gently brush them off or wait for them to fly away. They're attracted by perfumes we use, including those in sunscreen, shampoo, deodorant, and hair gels. Brightly colored clothing also attracts them. Yellow jackets need water and may be lurking in wet clothing or towels or around the pool, so watch where you sit. Yellow jackets cause the biggest problems when there's food around.

The Flying Insects

Before officials at Virginia's Great Falls National Park started managing the yellow jacket problem, visitors suffered about a thousand stings a year. Now, instead of serving sweet drinks in open containers, food concessionaires add lids and straws. They also use closed garbage containers, and they pick up the trash more frequently. Stings have dropped to forty incidents a year.

If you must eat outdoors, try one of the yellow jacket traps listed in the resource section. You can also attract them to plain old flypaper if you add attractive bait: dog food, ham, and other meat scraps in the early summer; sugary foods such as jam, syrup, and fruit later in the season.

Destroying yellow jackets is dangerous business, and something best left to professionals who have the proper gear. But if you have skunks or raccoons around your house, you'll be interested in this technique. A *Common Sense Pest Control* newsletter reported a way to get someone else to do the dirty work. One evening a naturalist with the East Bay Regional Park System in Oakland, California, poured honey over the entrances to ground wasp nests. The next morning, the wasp nests were all excavated and destroyed. It seems that many animal species were only too happy to dig up the nests to get the honey in the larval chambers.

FLOUR MOTHS

Have you ever opened a tightly sealed jar of flour and found it was full of wispy strands and weevils? It almost seems as if spontaneous regeneration had taken place. Where did these little critters come from?

These tiny specs in your food are actually the larvae of the flour moth, also known as Indian grain moths. You can recognize them by their color; usually they're white or yellow and have brown heads. The small brown moths that you see flitting around your pantry will sometimes lay their eggs in your dry goods, and soon these little weevils will hatch there. The full adult moths are slightly less than an inch long, usually gray and white with a fine powder on their wings.

If you're not careful, flour moths can spread from one product to another, and pretty soon everything will be infested. On a larger scale, this can be a big problem if you operate a flour or grain mill. The females can lay up to five hundred or more eggs at a time, meaning the numbers can grow at a frightening pace. When the temperatures are right—usually in the eighty-degree range—the eggs can hatch within three days. That's when you'll see the loose webs being spun. There's no comfort in knowing that this life cycle is continuous throughout the year. Every season is flour moth season.

Flour moths have very specific targets. In addition to flour, they go after bran, grains, all types of cornmeal and soy meal, dried fruits, nuts, cereals, livestock feed, and even pet food. If there are products that contain a high concentration of these foods, they will be targets, too. The actual culprits are the egg and larval stages; the adults don't eat the grains. They only live long enough to mate and reproduce. So the moths aren't the real problem. The challenge is to get to the early life stages of this pest.

One radical solution is to throw everything out and get new dry products. But first you should take the foods—in their containers—and put them in the freezer for four to six days. This will kill all of the eggs and larvae. After you then get rid of all the food, scrub all of the food containers with a strong solution of soap and hot water. Wash down all of the storage spaces and pantry shelves, too, because they may also contain eggs.

If you freeze your new flour, millet, or other dry goods for a few days before putting them on your shelves, you may prevent a new infestation. The freezing kills the weevils in the food. If you want constant shelf policing, try some flour bug traps. A new twist on old flypaper, the pheromone lures will attract and trap anything that likes to attack your flour, grains, dried fruit, dried milk, Dutch cocoa, and other cabinet staples. These lures, available through a number of organic garden suppliers, are small, rectangular boxes filled with sex pheromone. It's a powerful attractor for insects. The inside of the box, however, is lined with a sticky surface that traps the bugs. The lure works for about twenty weeks before you have to replace it.

Meanwhile, it's always a good idea to use containers that have tight-fitting lids. Food that comes packaged in cardboard or even thin plastic is better placed in a sealed container, especially in the warm summer months. You might even want to store your cereal, flour, and grains in the refrigerator if you're overly concerned about the problem.

APHIDS

If you see curled or wilting plants in your garden, there's a good chance that members of the aphid family have moved in. This is a very large group of insects, some of them winged and some wingless, which feed on the leaves, flowers, and stems of your cultivated plants. They are relatively small, anywhere from three to eight millimeters long, with a soft, pear-shaped body. Aphids come in many colors, from black to green and even pink. In spite of their sometimes-pretty colors, these critters are a serious pest if you are a garden lover. Aphids are also loathed because some are carriers of plant diseases.

Aphids live on the sap from plants. That's the same as sucking out their blood, and it's what causes the leaves to shrivel or wilt, and sometimes to drop right off.

CONTROLLING APHIDS

- Aphids don't generally like soil that is organically rich. If you can increase the mineral compounds in the garden, it will help to discourage their visits.
- You can make a spray from organic material such as cayenne pepper, garlic, and red pepper. Add it to a light corn oil and then spray it on your affected plants. (It helps to spray the plants with water first, to soften up the tissue.)
- Some gardeners also introduce "good" bugs into the garden to combat aphids. A few of the species you might try include the all-purpose ladybug, lacewings, and little aphid midges that feed on the pests.
- There's also a trend toward "companion" gardening; planting things that help and support each other. One of the better aphid defenders is the nasturtium, although it is sometimes the specific target of the black aphid.

In turn, the aphids secrete a kind of sap themselves, which can become moldy and spread diseases. It's actually called honeydew. This attracts ants and other insects to the same plants. The ants even cooperate with aphids and sometimes store their eggs and transport them to other plants. Meanwhile, all you see is a mass of little crawling creatures destroying your garden.

You'll find aphids all over North America, and more specifically on your prized roses, tomatoes, peas, beans, cabbage, fruit trees, and shade trees. There are even "specialized" varieties that favor apple trees, pine and spruce trees, and grapevines. The cotton aphid can be a particularly nasty pest for farmers.

GRASSHOPPERS

There are a great many varieties of grasshoppers throughout the world. The long-horned variety belong to the family Tettigoniidae and the short-horned to the family Acrididac. Most have two pairs of wings, mouths that chew as opposed to suck, and they like to eat plants. That's where we have the problem.

Grasshoppers have three life cycles: the egg, nymph, and the adult. Only the adult has wings and is able to fly. The female lays her eggs in the soil, sometimes she deposits as many as three or four hundred each summer and fall. Adult grasshoppers die at the end of each year, and the young survive any cold weather to hatch in spring. After they quickly become adults, they can also start laying eggs within one or two weeks.

A few of the more common varieties in North America prefer cultivated plants. When their numbers get large (as they do periodically), they can create a virtual plague. If they decide to migrate long distances, they can leave a path of destruction similar to a tornado. Grasshopper outbreaks have been recorded three times in the Great Plains, once in the 1870s, again in the 1930s, and more recently in the 1950s. These plagues usually coincide with times of terrible drought. It's almost as if the grasshoppers turn into marauding armies in tough times, scouring the countryside for food.

The members of the short-horned variety include the worst villains. One of these grasshoppers, the spur-throated, can swarm in numbers that exceed several million. If your crop is in their way, they'll simply eat their way through it. Normally the grasshopper will just eat leaves, occasionally moving down to stems, blossoms, seeds, or fruits. But when they move in hordes, they often eat right down to the crown of perennial plants. That means they get damaged beyond recovery, and they're ruined for the year.

There are certain field crops in which four of the varieties really thrive. About 90 percent of all the cultivated crop damage is done by the migratory, differential, two-striped, and red-legged varieties. They thrive on alfalfa, corn, soybeans, and wheat. There are at least ten potential insecticides that are registered for use on

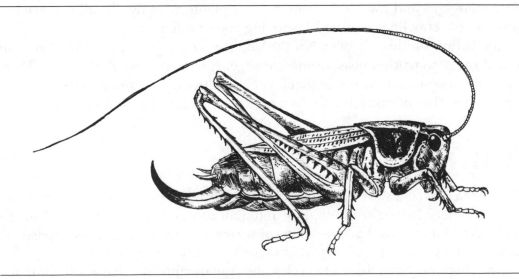

Long-horned grasshopper (Tettigoniidae)

crops, and several more (including carbaryl, diazinon, and acephate) that are approved for use in gardens or yards.

In May or June, grasshoppers are still in their young and immature stage. At that point, many die off during bad weather, or are eaten by predators or parasites. The worst time of year for grasshoppers is usually in July and August. You may find, for example, that when field crops are harvested nearby, these pests will move into your yard or garden and start to munch on your flowers or vegetables. Lettuce, corn, and beans all seem to be favorites. You may also see them on your shrubs, or attacking your fruit trees or shrubs.

One of the simplest steps to deal with a serious grasshopper problem is simply to plant vegetables that mature early. There is also some evidence that they don't like certain kinds of vegetables as much as others. Tomatoes and squash appear to be two of their least favorites. Some nursery owners may encourage you to use screens or fabrics to cover your vulnerable plants, but that's questionable advice. Grasshoppers have very strong jaws (called mandibles), and they can bite through some pretty tough material.

Researchers have found that there are several natural enemies of pest grasshoppers, and they can be introduced or encouraged. "A few of the most common and effective predators of grasshoppers include robber flies, spiders, and blister beetle larvae," say G. L. Hein and J. B. Campbell, entomologists at the University of Indiana. But according to Hein and Campbell, there's an even easier way to keep the grasshoppers at bay. "The most effective and practical natural enemy for use in yards and gardens is poultry, especially guinea hens." Apparently these birds can really rip into a plague of grasshoppers. Frogs, toads, birds, cats, blister beetles, and spiders can also help you win the battle.

Although chemical control is one option for winning the grasshopper battle, it's tough to beat out these large insects. You have to focus on certain points in their life cycle if you want to do any serious damage control. The best advice is to work on the adjacent places where they may be breeding; then move your battle into the yard. Start with the grasses and ditches in the areas around your property.

FLEAS

These annoying pests actually have no wings, but we'll deal with them in this chapter anyway because many people think they are flying creatures. "Flea" is the common name for any of the small wingless insects of the order Siphonaptera. Because blood makes up the complete diet of adults of both sexes, they are external parasites of mammals and birds. They have hard bodies, flattened from side to side, with mouthparts capable of both piercing and sucking. Their powerful legs enable them to jump high and fast (obviously the reason they are mistaken for flying creatures), which frustrates the hope of many a would-be flea killer. They are also hardy critters—adult fleas can survive away from a host for several weeks without eating. What is worse, fleas are often not specific to a particular host species, which means that the same one could jump from a cat to a dog to a person. Certain rat fleas transmit typhus and bubonic plague; rabbit fleas can transmit tularemia. Fleas also transmit tapeworms that can infect humans.

Fleas are sneaky little critters, too. They hitchhike into your house on your animals (and sometimes on you) and then move in. They make a home in your carpet and furniture, reproduce, and then plague you and your pets. So if you only treat your pets for fleas, without treating the house, too, they may still be reinfected later from the resident ones. You may first notice fleas when your pet starts to scratch a lot. (They're most prevalent during the warm spring and summer months.) You may also notice red spots (flea bites) and black-and-white pellets (fecal matter and eggs). Check your pet regularly for tiny black specs next to the skin.

As previously noted, fleas may not have any wings, but they can certainly jump, which anyone who has ever tried to pick a flea off a pet will know all too well. Just when you're sure you've smashed the little critter between your fingers, it leaps away. The best way to kill the fleas you do manage to pick off your pet is to fling them into a solution of soapy water or alcohol. During flea season, groom your pet regularly with a fine-tooth flea comb (fleas will stick to it if you put a bit of Vaseline at the base of the comb, and the Vaseline won't hurt your pet).

If your pet has more than a few fleas, you'll want to treat its entire body. Avoid the most noxious compounds and stick to insecticidal soaps or shampoos, or try dip treatments with earth or silica gel in them. Some ingredients you should avoid include piperonyl butoxide or petroleum distillates. Always read labels: What is safe and effective for a dog may be dangerous to use on a cat, for example.

Botanical insecticides like pyrethrum or rotenone are very effective both on pets and around the house. There are a variety of pet shampoos that combat fleas yet don't use harmful chemicals; they contain insecticidal soaps, oil of citronella, cedar wood, and eucalyptus—ingredients often found in natural insect repellents. Herbal flea collars are also impregnated with similar oils. Garlic and nutritional yeast added to your pet's diet may help it repel fleas; you can buy the latter in a prepared form with additional herbs, minerals, and vitamins added, or you can simply add yeast and garlic to your pet's food. Ask your veterinarian for advice on how much to use.

You need to treat the house, too, so wash your pet's bedding if you can. There are insecticidal soaps for use around the house. Two growth regulators—hydroprene and precor—prevent fleas from reaching adulthood, and both are effec-

CONTROLLING FLEAS

- Treat your infected pets.
- Clean up any of the affected areas in your home.
- Spray or dust the house for any additional flea problems.

tive tools for combating a flea infestation. There's even an electronic flea trap that attracts fleas from any room to a warm area covered with sticky paper.

While ultrasound devices don't work well on most insects, they do seem to affect adult fleas. Says Lowell Robertson, president of Sonic Technology, "Sound seems to interfere with fleas' ability to source the blood host. Our field-test trials show you can effect a substantial reduction of the flea population." These devices, however, cannot stop the animals from getting fleas on them when they go outside, he points out. An ultrasonic flea collar won't work because the sounds will be stopped or deflected by the animal's body. Save your money for a new leash or a scratching post. You might have to get rid of your dog's favorite blanket in the process, but she'll thank you in the long run.

3

THE RODENTS AND DIGGERS

The average homeowner would probably not use the word "hate" to describe how they feel about some of the common rodents. Squirrels and chipmunks can be nuisances, if they keep raiding your bird feeder, or stealing your cat's kibble. But you might use the words "fear" and "loathing" when your basement is invaded by them, or if your crops were suddenly overrun with four-legged gray critters. No one wants to share the breakfast cereal in their cupboard with a mouse—or their corn feed with a rat. Or their prized garden plants with a vole.

These rodents and diggers are industrious, clever, and persistent, so it's a challenge to find them and root them out. It also helps to break down the critters into their family groups, before we determine how to outwit them. Like the other pests we've examined, it takes a combination of tools, patience, and some basic science.

MOLES

Some of the best tunnelers in the mammal world are the little moles that you'll find across much of North America. You might even say that moles have tunnel vision. They spend most of their time underground hanging out in their well-constructed labyrinths. There they mate, raise families, eat, fight, and sleep. In fact, one of the few reasons they ever go aboveground is to find a new and better place to build *more* tunnels!

Nature designed the mole with this mission in mind. First, they are fairly small, between four and six inches long, with large claws and noses. You'll probably identify the mole by its large, pink proboscis, a tender appendage they never use as a digging tool. Instead, they do their earth moving with long, strong claws. Their eyes and ears (which are really internal) are hardly noticeable.

Some people get the mole confused with its cousin, the shrew. Here's how to tell them apart:

- The easiest way to tell them apart is their relative sizes. While moles are about the size of hamsters, shrews are the size of mice.
- Both mammals have a long snout, but the mole's is mostly pink skin. The shrew has a furry snout.
- Moles have stubby tails, which might be hairy or hairless. Shrews have longer and thinner tails with hair.
- The mole's front feet are noticeably larger than the shrew's feet.

There are seven species of moles in North America, although the most common ones are the eastern mole (*Scalopus aquaticus*) and the star-nosed mole (*Condylura cristata*). While both are found in the eastern half of the continent, the eastern mole is especially prevalent in the Midwest, the East, and the Southeast, while the star-nosed mole is found more often around the Great Lakes. A third variety, the broad-footed mole, is common in California. The eastern mole can be gray-colored in the north, and sometimes darker brown in other regions. They like grassy habitats, especially if they are moist and sandy. The star-nosed mole tends to be a bit darker, and it likes soil that is damp and near water. In fact, this critter is equally at home on land and in the water. You'll probably find the broad-footed mole in meadows and forests.

What all moles have in common is tunneling; they rarely come to the surface because they live quiet, solitary lives underground. Moles may share runway networks, especially if their home ranges overlap, but they don't socialize—they will fight when they encounter one another, sometimes to the death. Most moles are only sociable during the mating season in the spring. At other times they are vicious fighters, using their razor-sharp claws as weapons.

Moles tunnel underground because that's where their food is often found. Usually, their pathways are about ten inches below the surface. After a heavy rainfall, the eastern mole likes to forage. It eats a lot and so must travel far and wide to get all the edibles it needs. Moles feed on insects (and insect larvae), grubs, earthworms, carrion, the occasional unwary mouse (rodents sometimes use mole tunnels as shortcuts), and maybe even other moles they find in their tunnels. When underground food is scarce, they'll hunt aboveground, catching frogs and small mice.

Moles live a fast-paced and even frenetic lifestyle. Scientists have filmed them working underground, where they can excavate a tunnel at the rate of a foot per minute. They use their claws and webbed toes—with palms that face outward—as furious digging machines, literally "paddling" their way through the soil. As a result of all this work, they have enormous appetites. A mole eats between 70 and 100 percent of its body weight each day. A captive mole will actually starve in a few hours unless you keep feeding it. And if it eats more than it needs on any given day, the mole simply stores the food for the next day as excess fat in its tail.

Most moles have poor eyesight, and some even have their eyes covered by a thin layer of skin. Yet, that's not much of a handicap when you spend most of your time underground. To make up for this handicap, a mole has a terrific sense of hearing, touch, and smell. The broad-footed mole can actually track down an earthworm just by sensing its vibrations in the ground. The star-nosed mole, with its unusual nose, has twenty-two little "feelers" that can sense anything in its way—particularly food.

What makes these little critters particularly vicious is their set of sharp incisor teeth. They use these razor-sharp teeth to catch and hold onto their underground prey. Although they will occasionally eat plants and other vegetable matter, they usually leave these for the voles and the pocket gophers.

Moles produce two types of tunnels: some of them close to the surface that we see, and other tunnels deep underground. Moles make surface tunnels by pushing the soil up to form a roof, rather than pushing it out the end. During wet weather, surface tunnels are very shallow, but when it's dry, the moles burrow deeper for food. They dig the surface ones when they're hunting insects and worms. These tunnels often connect to deeper ones that lead to home chambers. Moles keep nests in these deeper and safer tunnels for resting and raising their families. The homes are in high and dry spots, perhaps under buildings or sidewalks. In order to conserve energy, tunnels may be used by successive generations of moles, and a mole continues to use the resources it creates.

Deep tunnels may be six to twenty-four inches below the surface. With so much heavy earth above these runways, it's sometimes impossible for the moles to create the tunnels by pushing the "ceiling" up. So, after loosening the dirt, the moles may push it back with their hind feet through the completed section and back to the surface or into an old shallow runway. As the mole makes the tunnel longer, it creates new exits so it doesn't have to push the excess dirt quite so far.

Mole holes usually open straight upward, unlike gopher holes, so the molehill will be a cone-shaped mound of earth. The gopher will often have a hole opening that is on the side of a hill, so its mound will be more in the shape of a fan. A molehill will certainly be smaller in overall size, too. These little excavators are at work all year, because they do not hibernate. They are most active, however, in the warm and wet months of the year. Those times often coincide with our own interest in lovely gardens and lawns. Moles thrive on our manicured lawns and golf courses, especially in the shaded areas. As a result, most of the varieties don't like to live in dry, semi-arid zones.

The star-nosed mole is actually at home on land as well as in water. You might find them near a pond, marsh, or stream, and you might even be able to trace their tunnels to these locales. This variety uses its large front feet to paddle through the water, and its tail becomes a kind of rudder for steering. The star-nosed mole has a slightly broader diet; it will also eat snails, bite-sized fish, cray-fish, and insects that thrive in the water.

Moles usually mate once a year. The female raises the young moles in her den, which is lined with grasses and other nesting material gathered aboveground. She may prepare several nests, some of which can be aboveground under a log or in a pile of vegetation. After a gestation period of four to six weeks, usually four young are born. They are helpless, furless, and sightless at birth, but they are nearly adult-sized and can see after just three weeks.

After about seven weeks they are on their own and must find their own home ranges. Often they will travel aboveground, one of the few times they'll be found there, until they find a suitable spot. Larger mammals such as coyotes, foxes, dogs, badgers, and skunks will dig out moles, and other animals may eat a mole that ventures aboveground. If they are wary, they may live about three years in the wild.

Golfers don't appreciate moles adding extra holes to their already difficult courses, so they have a gripe against moles. Many homeowners—at least the ones who prize a level and manicured lawn—don't like moles either. Or they think they don't. In fact, because moles eat insects such as Japanese beetles and grubs, and they aerate the soil during their tunneling, they shouldn't be considered such a terrible pest. They carry humus deep into the soil, and they bring subsoil close to the surface.

Homeowners also falsely accuse moles of causing damage to flowers, gardens, and landscape plants. Moles don't particularly like plants; any plant damage is more likely the work of voracious vegetarians such as voles, or perhaps mice. Niles Kinerk, director of the Gardens Alive! Natural Gardening Research Center in Indiana, says, "I don't mind them myself. Moles are kind of beneficial." But he understands how some people might think otherwise. "You run into these people who spend seven hundred dollars a year to have their lawn fertilized and mani-cured and have special lawn mowers that cut it in a special direction. Some people are pretty obsessive about their lawns," he notes.

Moles don't usually arrive en masse to your area. There tends to be a ratio of no more than two or three per acre. They like some protection, such as trees, side-walks, or buildings. And they really prefer some shade, too. If the surface area where they are feeding is cool as well, then the mole is likely to set up permanent residence. In spite of the good work they do as nature's bug eaters, soil aerators, and earthmovers, they can create quite a mess of an otherwise smooth lawn or garden. There's also some evidence that the tunnels they create may encourage other pesky creatures, such as mice and voles. In that way, these moles may be parties to other plant and crop damage.

There are several things you can do to encourage moles to do their tunneling elsewhere, Kinerk explains. You can apply milky spore or beneficial nematodes to your lawn; they kill the grubs, a main food source for moles. Nevertheless, the problem may get worse before it gets better, as the hungry moles frantically search the area for food before heading to greener pastures.

Another approach for a relatively small area is simply to water it thoroughly. Getting the ground wet once or twice a week for three or four weeks may deter them, says Kinerk. Alternatively, you can try drying out the soil and forcing the insectivorous mammals elsewhere. Packing the soil with a roller may also make it too compact.

Repellents may work, too. Mothballs and flakes, of course, can discourage the diggers. An old folk remedy—planting the castor bean or caper spurge—may repel the animals, too. Remember, though, other animals and children may try to eat the mothballs you spread in your yard. A better technique would be to drop them down the mole holes. As for electromagnetic repellers, which are touted as a solution, once again you'd be best advised to save your money. There's very little scientific proof that they have any effect on moles.

Another solution is the stink-em-out routine. Sprinkle a castor oil concoction on the ground. Whip two ounces of castor oil with one ounce of liquid dish detergent in a blender until it holds its shape. Add water equal to the volume in the blender and whip it all again. Fill a sprinkling can with water, add two tablespoons of the castor oil mixture, and then sprinkle on areas of heaviest burrowing concentration. This works best after a good rain or watering.

Vibrations can also scare away moles. Try pushing a child's pinwheel down in the earth in several spots in the area where the tunnels are dug. Or you can purchase a commercial mole windmill or vibrator like Plow and Hearth's Mole Evictor. Each device is effective for five hundred square yards, and you can choose between solar-powered or battery-powered machines. Unlike the children's pinwheels, these work without wind and send out mole-annoying sound waves from their buried aluminum shaft.

Trapping is a popular method of controlling mole problems. One of the recommended varieties is the Victor, which is a kind of "harpoon" trap. Another brand is the Out O' Sight trap, which uses a scissor-jaw device to kill the mole.

TOTAL CRITTER CONTROL

And the third variety is the Nash trap, which uses a loop device to choke the mole. If you're worried about the pain inflicted by these traps, they all use the "quick-kill" method.

You'll need to do a little investigating before you set your traps. They will work best if you place them in the surface runs, because these are probably the areas of most activity. If you're not sure if a mounded run is in use, just gently push it down with your foot. Moles are pretty conscientious, and they'll be along that same day to fix the tunnel you've damaged. Then you'll know that you can confidently trap them in that spot.

Apparently moles are pretty smart; they can actually detect a trap that's been poorly set. So you'll need to follow the instructions carefully, noting the right level at which to set the device in order to catch the mole on the move. If the mole detects the trap, it will simply burrow under or around it. For that reason, it's not worth leaving a trap in one place for more than two or three days if you don't see any action.

It's a common belief that grubs bring the moles. Although that's partly true, a mole eats many kinds of bugs and underground food. You can try to rid your lawn of grubs by drowning them, but that might bring the moles just so they can get some moisture. If you try insecticides in an effort to get rid of their food, you might simply kill all of the other underground life—much of it helpful to your lawn or garden.

Repellents are worth a try, and scientists at the University of Nebraska at Lincoln have had some success in this area. Dennis Ferrara and Scott Hygnstrom, who specialize in wildlife damage control, have found that castor oil and a castor oil product named Mole-Med have shown pretty favorable results. "In one study," they note, "Mole-Med successfully reduced mole activity in an area for over thirty days." To get those kinds of results, you'd have to really work the castor oil into your lawn. You'll have to water extensively, then add the repellent, and then water the area again. And you can't just water the perimeter; the moles will simply tunnel under it.

Ferrara and Hygnstrom offer a cheaper alternative, too. "Homeowners can prepare their own repellent concentrate by mixing six ounces of 100 percent unrefined castor oil (refined is suitable, but expensive) with two tablespoons of liquid detergent in one gallon of water." If you dilute this mixture at the rate of one ounce per gallon of water and then apply it liberally with a sprayer, you should get good coverage on about three hundred square feet of lawn or garden.

Some people get desperate about moles and turn to fumigants. These come in little gas cartridges, frequently using carbon monoxide or carbon dioxide. Farmers sometimes turn to stronger materials, such as aluminum phosphide (a pesticide). With fumigants, you need to get the cartridges right down into the deepest runways, and then seal up all of the possible openings.

In a few parts of North America, you're allowed to use toxic baits to kill moles. One of the most common varieties is strychnine; it's a federally registered pesticide for this critter. As researchers have pointed out, however, moles prefer to eat worms and insects, so they aren't too likely to take a bait. You also run the risk of poisoning other creatures at the same time.

Mole barriers can be a lot of work, but they have proven to be effective. They involve the use of aluminum sheeting that's at least a yard wide, or galvanized hardware mesh cloth that has openings of no more than a quarter inch. You then have to bury this "wall" up to twenty-five to thirty inches in the soil, while leaving it about six inches above the ground level, too. This method is obviously only practical for a small, enclosed space. It also involves a lot of money and expense.

As with any pest problem, there are a whole slew of home remedies that are offered as the miracle solution. Gardening with gopher purge (a yard plant) has been suggested, but there's little or no proof that it actually keeps away pests. It's also a poison to humans, and can become a problem weed itself. That's one "solution" that you can scratch. Others that have been suggested are: windmills, rose thorns, laying out broken glass, used kitty litter, even flooding the ground with kerosene or bits of chewing gum. As you can see, some are pretty outrageous; most are simply unproven.

Another approach, one that your kids will probably enjoy, is to walk on the tunnels and simply collapse them. While this sounds cruel, you're unlikely to actually kill any moles because they live in the deeper tunnels. But if you stomp the tunnels frequently enough, the mole may get the idea that the area is inhospitable and abandon it.

Trapping is still probably the most successful approach. An infestation of moles would probably be three or four an acre, so you should be able to catch the animals. Some mole traps impale unsuspecting tunnelers as they go about the business of getting a little dinner. Others cut them in half or choke them. It's not a pretty business, but do not despair. You can also live-catch your moles.

Find an active mole runway and set up a "pit" trap. Dig into the runway and place a three-pound coffee can or wide-mouth jar in the dirt below the runway. Then cover your excavation with a board, making sure no light gets in. And wait (while checking twice a day). Once you have a captive mole, be sure to release it far away from your lawn.

Moles are one of the most abundant of small mammals. Their innocuous behavior, combined with their positive "ecological" role, actually makes them a boon. As long as you can overlook those little rises and furrows in your lawn.

TOTAL CRITTER CONTROL

VOLES

In the yard, moles and mice are often confused with voles, who are vegetarians and eat their weight in plant food every day. Voles (family Cricetidae) are stubby, stocky, brown or gray animals with dense fur and short legs and tails. Mice have longer tails, voles are stockier in build, and voles also have bigger eyes. They're about three to five inches long, and the tail will be anywhere from a half inch to an inch and a half long. Look for brown fur on their sides and back, and lighter brown to gray underneath.

There are three members of the vole family that annoy us—the meadow vole (*Microtus pennsylvanicus*), the prairie vole (*Microtus ochrogaster*), and the pine (or woodland) vole (*Microtus pinetorum*). The first two species are pretty common across most of North America; the pine vole prefers the broad-leaved forests of the East and the pines of the South. Their worst feature, from our point of view, is the rate at which they multiply. Each female vole can have up to a hundred offspring in one year. In years of "plagues," they can multiply from one hundred per acre all the way to ten thousand per acre.

These little critters can be found in forests, fields, orchards, or grasslands. They live outdoors in areas where they can find lots of cover, such as grassy or littered fields or orchards. They eat these grasses, as well as seeds, tubers, bulbs, roots, and occasionally snails or insects. In the fall and winter, they're especially likely to eat bark. Some voles will "girdle" trees, eating the bark around the entire circumference of the tree—which kills it. You'll see this damage in the winter when the voles may kill young shrubs and trees in your orchard, or along a windbreak in a field. Remember that they aren't hibernating; they're simply busy burrowing along in snow tunnels just above the grass.

The pine voles, meanwhile, are busy eating the roots and tubers of trees and plants. According to researchers, they'll also eat the leaves, shoots, and seeds of grasses and broad-leaved flowering plants. Flower bulbs, vegetables, field crops, and lawns may all fall victim to this little critter. You can recognize their telltale sign by the tooth marks they leave on your flora. The individual marks are about one-sixteenth of an inch wide (rabbits will be about twice that size), they'll be irregular, and at a variety of angles on the stem or bark.

You'll be able to spot the prairie and meadow voles by their "runways." These are little one to two inch paths on the surface of the ground where the vegetation will be close-cropped and may, in places, have a little canopy over it. You may also spot a little entrance to an underground home at the end of the runway, plus a bit of litter from leftover food. The pine voles are harder to spot, because they usually travel in underground tunnels and only spend a limited amount of time aboveground. The biggest problem from an agricultural standpoint is their love

The Rodents and Diggers

Meadow vole (*Microtus pennsylvanicus*)

of hay, sugar beets, and other field crops. In some areas they will also attack peanut crops and even tomatoes.

Their home range is usually a quarter acre or less, depending on the population density, habitat, and food supply. They move around in a complex tunnel and surface runway system, often used by several adults and their young. Vole runways are about an inch or two wide and are the most readily identifiable sign of voles. Remember that they're active day and night, eating and moving around.

Voles may breed year-round, but tend to multiply mostly in the spring and summer. They have five to ten litters a year and produce three to six young each time. Young are weaned after twenty-one days, and females can breed at thirty-five to forty days old. Scientists were shocked to find that one pair of voles held in captivity had seventeen litters in a single year, producing an astonishing eighty-three young. One of the females from her first litter had thirteen litters herself, producing seventy-eight young in her first year.

Fortunately, voles are an important source of food for many other creatures. Snakes, hawks, badgers, mink, foxes, martens, coyotes, and owls all feast on these little critters. They don't live for long, a price to pay for their frenetic lifestyle, and an old vole is sixteen months. Most voles live, on average, about two or three months. There are cycles, however, so you'll notice the population rise and fall dramatically in some years.

They don't usually enter our houses, but in some rural areas they may be a problem for farmers or homeowners. You may first notice voles around your house when their population peaks, as it does every two or three years. For a homeowner, dealing with voles should be fairly easy. If they're bothering a few trees or seedlings, you can protect them with quarter-inch hardware cloth cylinders buried about six inches into the soil. Unlike mice, voles pose few health problems for humans, although they occasionally carry some diseases: plague and tularemia, for instance. So think twice if you're ever tempted to pick one up.

You can generally get rid of voles by altering their habitat. Eliminate weeds, ground cover, and litter that provide food and cover. Mow the lawn regularly and keep mulch about a yard away from tree trunks so they can't get close under the cover and eat. Tilling the soil will destroy the runway and tunnel system, and probably reduce the population somewhat. You could even burn the ground where they are building their runways. And, as with other rodents, you can protect some of your gardens, flowerbeds, or vulnerable trees by using woven wire or hardware cloth fences. You'll need to make the barrier about twelve inches high, while the bottoms should be buried slightly or at least fit tight to the ground. With the pine voles, of course, you'll have to go at least six to eight inches below the ground.

Remember that you'll have to adjust these exclusion devices higher if there is snow in your area. You also need to plan ahead when you are installing cylinder devices. Anticipate how much the plant or tree will grow, and then add in that much space for the device. Depending on the weather in your area, these devices should last four to six years.

As with moles, traps and toxicants are also an option. The traps can be set in singles, facing across the animal's runway, or you can set double traps back to back to catch the voles from either direction. The basic snap trap that you'd use for a mouse or rat will work well. A common bait for them is peanut butter sprinkled with oatmeal. You could also make little homemade traps by dabbing glue inside black plastic piping. Fumigation doesn't work well with voles because their runways tend to be either aboveground or very shallow. Aluminum phosphide is again an option, but it always requires special care. Remember to read all of the labels carefully when you are dealing with these hazardous chemicals. If they are deadly to voles and moles, they most certainly aren't going to be healthy to you, either.

POCKET GOPHERS

Pocket gophers look like smaller versions of beavers, but most people will never get to see one since they spend most of their lives underground. Pocket gophers (family Geomyidae) get their name from the fur-lined pouches next to their mouths, which they use to carry food. They can also close their lips behind their front teeth (their four big incisors) to keep soil out of their mouths while burrowing. A distinguishing feature is the groove on the outside of the two front top teeth. In function, but not in family, pocket gophers are a close cousin to moles. Both species are born tunnelers.

Don't confuse pocket gophers with chipmunks. Those are the harmless little guys with the stripe down the back that scurry aboveground and feed on seeds and nuts. The destructive gophers are underground. Although gophers churn and aerate your soil, they aren't there for altruistic purposes. Gophers in your garden are after plants, and they are especially fond of bulbs.

Pocket gophers are small-eyed, short-legged, stout-bodied rodents, usually about six and a half to ten and a half inches long. They are frequently brown on their upper parts and grayish underneath, but their coloring can vary widely from region to region. Scientists have found almost-white pocket gophers in desert regions, and there are a few varieties along the Pacific Coast that are nearly black. (This is directly related to the nutrients and organic material found in the soil in which they live.) They weigh about a half to three-quarters of a pound, although the northern variety tends to be smaller than those found in the South. Softer and deeper soils tend to encourage the larger varieties. There are thirty-three different species, but the pocket gopher is found in some form or another across much of North America.

Besides their prominent front teeth, the pocket gopher has eighteen other teeth designed for chewing. But the real distinguishing feature of this mammal—and the thing that makes it also a relative of the ground squirrel—is its long front and back claws that it uses for digging. This industrious little animal can burrow tunnels up to eight hundred feet deep in the ground. But while their fellow diggers, the moles, are mostly beneficial, many people consider gophers downright destructive.

Larger than moles, pocket gophers dig bigger holes—making them unpopular with golf course owners. They are bulb- and root-eating vegetarians, so these critters are the nemesis of gardeners and farmers alike. And with their large incisor teeth that can cut through almost anything in their path, they've earned their share of enemies across North America. In Nebraska, for example, these little critters inflict more than $10 million in damage upon farmers' major field crops. If they settle into alfalfa fields, damp meadows, or cattle ranges, for example, they can cut the yields by 30 to 40 percent annually. Pocket gophers can also cause damage to fruit orchards, windbreaks on farms, public parks, golf courses, and even the runways of airports. Gophers can even chew through metal, including underground pipes. And they can sever electrical and irrigation lines that happen to be in the path of their tunnels, too.

These critters managed to raise the ire (not to mention the roof) of at least one family in Minnesota. It seems that back in 1989, a gopher chewed through a thick polyethylene gas pipe leading into a home in St. Cloud. Apparently the gas accumulated and ignited, blowing the house up! It's not surprising that some irate citizens encourage flooding of affected areas to get rid of these critters.

Gophers have strong shoulders and long claws on their forepaws for digging. Because they spend most of their time in tunnels, their whiskers help them to sense their way around in the darkness, and their tails help guide them when they go in reverse. The gopher rarely ventures aboveground, except perhaps to gather bark, grass, or greens, which the animal stuffs into its cheek pouches for later feeding underground. Usually they are only seen when they push soil to the surface, or if they are on the move to a new habitat. But most pocket gophers stay in the same set of tunnels for their entire lives—which usually only last about two years.

The gopher's life is a subterranean one. An efficient digger, a single gopher can tunnel two or three hundred feet in a night. In fact, each pocket gopher is capable of moving up to two tons of soil a year! All of this dirt ends up in mounds (about 150 mounds per year per gopher) piled next to their holes. Like other underground diggers, each adult usually has its own set of tunnels throughout the year, and they tend to make most of their mounds in the spring and fall. (Unlike moles, they don't tunnel close to the surface and push up the soil.) When they find a good source of underground food, they'll carry it to a storage area that is usually sealed off from the main tunnel.

Why all this emphasis on earth moving? Gophers tunnel beneath the surface of the soil, eating the roots and tubers they encounter in the process, pulling the vegetation into the tunnel from below. They also like to eat crop plants, bulbs, shrubs, grasses, carrots and vegetables, nuts, and broad-leafed weeds. They'll strip the bark off young trees right at the ground level, and cut off their roots as well. Their burrowing holes and mounds can also damage harvesting equipment, or injure cattle and horses.

Gophers also use tunnels for breeding, nesting, and resting places. Since gophers don't hibernate, tunnels also have to be deep and warm, although some northern gophers will tunnel in the snow and pack the tunnels with earth. When the snow melts, the earth tubes remain aboveground until they disintegrate. Burrow systems may be straight tunnels or branched labyrinths. How far the gopher tunnels depends on how much food is in the area. Solitary and antisocial by nature, one gopher lives alone in each tunnel, unless mating or caring for young.

In the northern parts of their range, gophers breed once a year. In the irrigated fields of California, however, they may breed year-round. The female births from one to seven young, and in some species, the male may stay around long enough to help care for them. Once the young are weaned, they leave and begin to dig their own burrows. They're sexually mature at a little under a year old. The young are usually born between October and June.

The Rodents and Diggers

Pocket gophers have plenty of enemies. Weasels, skunks, and larger snakes all come into tunnels to find them. Other predators dig them out—badgers, for instance. Aboveground, a world of predators—raptors, coyotes, foxes, skunks, bobcats, and even housecats—all wait for a gopher to deposit a little soil on the mound or venture out for a bite of grass. Owls will also happily take an unwary pocket gopher if it lingers too long at the mouth of a tunnel.

Even with all the pocket gopher's natural enemies, more than enough of the species survive to wreak havoc in your garden. One effective, if labor-intensive, way to protect your plants is with root guards, according to Mark Fenton of Peaceful Gardens, an organic garden supplier in California. Root guards, which you can make or buy ready-made, form a protective cage around a plant's roots. Fenton says you can use chicken wire to make your own root-guard baskets, but small gophers can get through even one-inch or three-quarter-inch holes. If you want the greater protection afforded by half-inch or smaller holes, use aviary wire, which has a finer mesh.

Some people use root guards for flower bulbs; others for ornamental or fruit trees, says Fenton. Obviously it's much too labor-intensive for vegetable gardens. Tree roots will be able to get through the wire as the tree grows, although this will eventually cause the guard to become ineffective.

An easier solution, however, is to trap the gophers. Like most mole traps, gopher traps are deadly. Gopher tunnels are a little more difficult to find than the surface mole tunnels. You have to dig a ring about a foot and a half away from the mound until you find the main tunnel. You could probably live-trap them in a large pit trap, catching them in a gallon-sized jar. Or, you could try getting a large cat. "There are a lot of cats that are good gopher cats, and they will keep yards clean," says Fenton. "It probably depends on how well they are fed." Terriers and some other breeds of small dogs are also ferocious gopher getters.

You can certainly try keeping the soil very wet. It's difficult for gophers to dig in very wet earth; it clumps on their claws and gets their fur dirty. In addition, wet soil will trap noxious gases in the tunnels with the gophers, while the dry soil naturally allows air exchange. Although it is effective, you can obviously only use this tactic on an area the size of a small garden, not an entire field.

Many companies sell earth vibrators, stakes that when implanted in the ground send out small shock waves that frighten away burrowing mammals. As with other gadgets sold for garden and yard use, it's hard to tell how well (or sometimes even why) they work. Fenton says that manufacturers claim the effectiveness of the device depends on where it's placed and on the composition of the soil (which affects how well vibrations travel through it). "I think it works not as well in some sandy soils, because the vibrations don't travel as far in sandy soils as they do in a heavier clay soil."

Some people also poison gophers. In fact, Nebraska's alfalfa farmers are so incensed about pocket gophers that many pour poison directly down the gopher holes. They do this with machines that are called "burrow builders." This equipment actually burrows down into the ground, intercepting the gophers' tunnels, and dropping in toxic baits. You need a tractor and some fairly sophisticated machinery, but there are experts who specialize in this sort of rodent control.

Some frustrated gardeners and farmers have tried mothballs or flakes as a repellent. Fortunately, technology may be coming to the rescue in this case. Researchers are breeding alfalfa plants that not only seem resistant to gopher teeth, but that are actually helped by gnawing. As a matter of fact, horticulturists are designing gopher-resistant plants. Perhaps scientists can even develop a tulip bulb that's gopher-friendly.

WOODCHUCKS

Almost everyone has heard the old saying, "How much wood would a woodchuck chuck if a woodchuck could chuck wood?" The real question should be, "How much earth could a woodchuck move if a woodchuck could move earth?" Since woodchucks *can* move earth, it's possible to get a pretty accurate answer. According to the U.S. Department of Agriculture Extension Service, a single woodchuck can move more than seven hundred pounds of earth in one day. Whether they do or not, that's anybody's guess.

Woodchucks are found in the Plains states and throughout much of the eastern United States and southern Canada. Also known as groundhogs, these animals have compact, chunky bodies sixteen to twenty inches long covered with a grizzled brownish gray fur and are supported by short, strong legs. Males weigh five to fourteen pounds; females are a little more petite. The scientific name for woodchucks is *Marmota monax*. *Marmota* places them in the squirrel family; *monax* is derived from the American Indian word for digger. And, true to their name, woodchucks are certainly diggers. They burrow into the earth to create their homes—tunnel and den systems where they hide when threatened, mate, wean their young, and, in winter, hibernate.

Woodchucks are specially equipped to do all of this dirty work. To dig up the earth they have short, powerful legs equipped with strong claws. Woodchucks push loosened earth out of the burrow with their blunt heads and chests. To keep dirt out of their ears—a problem if you happen to be an animal that pushes loose dirt around with your head—these earthmovers can actually close their ears.

Woodchucks prefer to live near open farmland. Their burrows are commonly found in fields, pastures, along fencerows, stone walls, roadsides, at the bases of trees, or adjacent to gardens. You can find active burrows in spring or summer by looking for large mounds of freshly excavated earth at their main entrances. They're usually about ten to twelve inches in diameter. In addition, there are usually between one and four escape holes. These secondary exits are harder to find because they are usually located in thick vegetation. They're also dug from below the ground, so they probably don't have mounds of earth beside them.

The length of a tunnel varies between eight to sixty-six feet, and leads eventually to a nesting chamber, which is about sixteen inches wide and twelve inches high. Woodchucks use this den for several seasons and hibernate in it until the late winter. When the males emerge, they fight each other for tunnel territories and to win over the females. Because they have such large and sharp teeth, these battles can be quite vicious.

Woodchucks spend all their energy on their tunnels for security reasons: They have plenty of enemies. Hawks, owls, foxes, bobcats, weasels, dogs, and humans are all chasing them. And because the woodchuck doesn't run all that quickly, it needs plenty of holes into which it can escape. In fact, a woodchuck usually ventures no farther than fifty to one hundred feet from its den in a typical day, although the distance may vary based on the availability of food.

When a woodchuck wants to see if it's safe to venture out, he pokes the top of his head over the rim of the burrow. Interestingly, woodchucks' eyes, ears, and nose are located toward the tops of their heads. This adaptation allows them to see if the coast is clear, while keeping most of their body hidden in the tunnel.

The main reason woodchucks leave the burrow is to eat. They are primarily vegetarians, as are many other rodents. Woodchucks have white, chisel-like incisor teeth, which they use to efficiently chomp away at the products of our fields and gardens. They feed aboveground, mostly in the morning and evening, enjoying a variety of wild grasses and field crops, such as alfalfa, clover, and legumes. They also like vegetables, including peas, beans, and carrot tops. The more tender and succulent, the better.

When not feeding, woodchucks sometimes spend the warmer part of the day basking in the sun. You might spot a woodchuck dozing on a fallen fence post, a low stone wall, large rocks, or a log pile. These periods of leisure, however, are usually spent close to the burrow entrance. If danger threatens, the woodchuck will scurry inside. If a woodchuck is startled, you might hear it emit a shrill whistle, or alarm. This is followed by a low, rapid warble that sounds like *tchuck, tchuck*—the sound that probably inspired their name.

As colder weather rolls around, woodchucks begin to hibernate. In fact, they are among the few mammals that enter into true hibernation. This usually occurs in late fall, near the end of October or early November, although the start of hibernation varies with the latitude. They continue to hibernate until late February or March, when they will emerge from their burrows to find a mate.

Breeding occurs in March and April. After a gestation period of about a month, the female gives birth to a single litter of two to six (usually four) young. The babies are born blind and hairless, and very small, but they are ready to emerge from the den within a month. In the meantime, both parents stand guard against predators. The babies are weaned by late June or early July, and soon thereafter strike out on their own. They frequently occupy abandoned dens.

Mature woodchucks will often use a burrow and den system for several seasons. They keep the burrows clean, and annually replace nest materials. When they abandon their homes, other animals—including rabbits, skunks, foxes, and weasels—may move in.

One benefit of all this digging is that it actually improves the soil. Woodchucks condition the soil through their burrowing activities. Nevertheless, it is hard to convince farmers and gardeners that these are benign creatures. The problem is that they like to eat some of the same foods we like to grow. They love gardens—and they're notorious for climbing fences to get into one. They may even gnaw on your pipes or wiring. Obviously you'll need ingenuity to get rid of these critters.

Start by making your garden harder for them to access. Mesh fences are a good alternative for your vegetable garden, if you're tired of doing all the work but reaping none of the results. These heavy-duty wire barriers will need to be at least three feet high and you might want to bend the fence outward at a 45-degree angle—woodchucks are good climbers. Bury the wire six to twelve inches below the ground as well, to prevent woodchucks from burrowing under it. To further discourage climbers, you might install an electric "hot-shot" wire four to five inches off the ground and the same distance outside the fence. (Some people have found that a hot-shot wire alone—even without a fence—will deter woodchucks from entering a garden.) Before installing the hot wire, remove any vegetation near it so the system won't short out.

Another way to keep unwanted burrowing animals out of your garden is to plant only things they don't like. Castor beans discourage gophers and moles, but castor beans are also poisonous so they're not good plants to have when children are around. Try different vegetables—eggplant, radicchio, or broccoli. Some gardeners find that woodchucks are selective and won't eat plants they don't like if more tempting food is available.

You can also try to simply chase the woodchucks out of your area. There are several different ways to do this. Because woodchucks are easily frightened, a large leashed dog will probably keep them away (although then you have a dog to care for). Where woodchucks are particularly wary, you might even find that used pie-plates strung up on a rope, or scarecrows placed in a garden, will be enough to discourage them.

Another approach is to find the woodchucks' burrow, and then make life unpleasant for the residents so they'll move of their own accord. You can determine which tunnels they're using by stomping down any mounded openings you find; the ones that are re-dug are active. Then you simply drop unpleasant things into the tunnel opening. Various authorities recommend shoveling in some dog droppings, or stuffing up the hole with a rag soaked in peanut or olive oil. (Apparently the oil becomes rancid and will stink the animals out.)

For more urgent woodchuck problems, you'll have to resort to tougher methods. One of the more common means of woodchuck control is to use a commercial gas cartridge. This is a specially designed cardboard cylinder filled with slow-burning chemicals. You ignite it and place it in the burrow system; then all the entrances are sealed. As the gas cartridge burns, it produces carbon monoxide, which accumulates in the burrow system and kills the woodchuck. Gas cartridges are available from farm supply stores or your local Fish and Wildlife Service.

You might also want to consider traps. If you use live traps, check them twice a day so the animals don't suffer while awaiting transport. Release them where they'll find plenty of wild food, such as grasses. Don't trap them in the late summer, before they hibernate. Wait until the early summer, after they've awakened from hibernation and are well fed and strong.

Kill traps are a more drastic solution. These are available as boxes or as leg-hold versions. They'll need to be placed strategically at an entrance, and you may even have to rig up some little stick guides to steer the woodchuck into the trap. Some of the baits that seem to work best are lettuce, carrots, or apple slices.

If you're getting desperate, you *might* also try shooting woodchucks. Apparently their meat is rather tasty, and in some regions you'll find that woodchucks are game animals (you'll need a proper license to hunt them). In other places they aren't governed by any hunting regulations. Check your local laws before you take the .22 rifle or the shotgun out in the backyard.

Finally, it is absolutely a myth that the groundhog comes out of its burrow in early February to check its shadow and help to "predict" the weather for us. Fortunately for all of us, the groundhog family is still fast asleep during their long winter's nap.

MICE

Many people seem to have a double standard when it comes to mice. In books and cartoons, mice are celebrated. There's Mickey Mouse, Minnie Mouse, Mighty

Mouse, and other pint-sized squeaky heroes. Even when it's cat-versus-mouse in the cartoons, the mouse is always the hero. And witness the many varieties of cute, stuffed, mouse toys for children. Why do we then go about trapping and killing them when they come into our homes?

Perhaps it's because mice like to eat your food before you do. They enjoy tearing apart pillows, mattresses, stacks of magazines, and other shredable objects to turn them into nests. They leave little droppings all over the place. And they carry diseases, including rat-bite fever, tapeworm, ringworm, leptospirosis, and salmonellosis. Leptospirosis, which is spread through urine and can cause kidney and liver damage, is rare, and rat-bite fever is even rarer. Salmonella and other microbial food poisoning is a greater possibility, with probably thousands of cases each year. You can get these diseases from mouse bites or by eating food that mice have been munching on—and contaminating through their poor table manners.

Common house mice (*Mus musculus*) are little rodents with small, black eyes and big ears. They're usually light or medium gray, with long tails measuring about three to four inches. The body of the adult is a bit less than half as long, so the tail is a reliable identifier. In spite of their size, house mice only weigh about half an ounce, and their bodies are surprisingly flexible. As you may have discovered, they can squeeze through the tiniest holes with no problem.

Mice are also destructive, which is why they're not hard to notice in the house. Maine resident Jane Connors was a typical victim of a mouse infestation. She first discovered the little critters snacking in her couch. Initially attracted by food crumbs dropped by snacking TV fans, the mice then decided to stay. They gnawed and burrowed their way into the furniture and made a serious nest of ac-

Mouse (*Mus musculus*)

cumulated foam rubber and fabric. Connors discovered the hole when she was looking for the TV remote control, but there are other telltale signs. They may leave little tracks, especially if they've been in wet or dirty areas. You'll notice four or five splayed toes, for instance, on the floor where they've traveled.

Other signs are gnawing marks on food containers. Their teeth are small, so these little bite-marks will be knife-thin. Mouse droppings are a definite calling card from your visitors; they'll be about a quarter-inch in size. If there are corners they're scooting around, or vertical surfaces in their nest area, you'll see rub marks on the walls, plus little piles of seed shells or other leftover foods they've brought back to the nest. You might also notice a musky smell where they're hiding.

House mice are generally nocturnal, although you may see one during the daytime. While mice have a good sense of hearing, taste, smell, and touch, they don't see very well and are color blind. They're good athletes and can generally do whatever gnawing, jumping, or swimming it takes to get where they're going. They're also excellent climbers, which means they can reach almost any surface. Don't be surprised if you spot one seven feet above the ground, running along the curtain rod. House mice are fast, agile, and apparently fearless. They can run up any rough vertical wall surface, and they can leap at least twelve inches while standing still. As for hole openings, they can fit through a space as narrow as a quarter inch.

There are more than 250 species of mice in North America. The white-footed mouse and deer mouse are the most common ones you'll see outside, although they'll come indoors, too, seeking food and warmth. Other species, such as the brush mouse, cotton mouse, and piñon mouse, for example, are more specialized and limited in range. The familiar visitor to our domiciles, the house mouse, is a different species and was introduced to the continent in the sixteenth century from Asia. Outside of rural areas, this is the mouse you'll see or hear in your house. You can tell them from white-footed and deer mice, because house mice are gray and have long, scaly tails, and the white-footed mice are brown to near black with white undersides and feet.

Despite the name, house mice may live outside of human homes, even in open fields. But they're most likely to be in the vicinity of homes and commercial buildings. (We'll include white-footed and deer mice in this section, because the general biology is similar.) Mice have small home ranges, and may only travel within a small woodland, barn, or home. Studies show that they usually settle within five hundred feet of their birthplaces. All mice enjoy a visit inside your house, especially when it's cold, and some obviously stay too long.

House mice eat nuts, seeds, fruits, berries, mushrooms, and insects—or frankly, whatever is in the food cabinet. White-footed and deer mice will also eat fruit, insects, fungi, and possibly some green vegetation. Even house mice prefer seeds and grain, but they are dedicated nibblers, trying everything just to see if they like it. And they probably will. Vulnerable pantry items include high-fat treats such as nuts and chocolate candies. It's not that they eat so much, just about three ounces a day, but they sample everything. They destroy and damage far more food than they consume, and house mice can get by on little or no free water because they get the water they need from the food they eat.

Since they don't hibernate, some house mice cache food for the winter, and they hoard more food in colder parts of North America than they do in the warmer regions. White-footed and deer mice will store food near their nest sites, especially in fall and winter. They don't necessarily eat the food they hoard, and the attractive caches lure in a host of other pests.

House mice generally breed year-round unless they live outside. Then, like the white-footed and deer mice, they breed in spring and fall. They can become pregnant early in life, sometimes within six to ten weeks. (While that seems premature, consider a house mouse's lifespan: nine to twelve months.) While all the jokes about prolific breeders focus on rabbits, a female mouse can have as many as eight litters a year! If they have five or six young each time, you can see how your house could be overrun with the little critters in no time.

The males and females may stay together for a few days or for the whole breeding season, but generally they take new mates after winter passes. The female usually has a twenty-two- to twenty-five-day gestation period before she gives birth in a nest she constructs to receive her young. The nest, a rough cylinder of shredded paper or other fibrous material about six inches in diameter, is usually stashed in a private crevice four to ten feet aboveground, such as a crack in a wall or a birdhouse—or it may be a hole in the ground. In a house, the nest could be in just about any sheltered location.

Baby mice, five or six of them, are born hairless and blind, but after two days they begin to sprout fur. After about a week, they have nearly all their fur, and after two weeks, they open their eyes. They're weaned around the third week, and shortly after, they begin to venture out of the nest. By six or seven weeks old, they're sexually mature and set out on their own. The female will breed again after the young leave, and will continue to raise four or more litters over the season. Mice usually live about a year in the wild, which is a long time considering the number of animals who eat them. Most predators will kill a mouse: foxes, raptors, blue jays, snakes, coyotes, and bobcats, to name only a few.

Getting rid of your own mice won't be easy. Mice like living with people, and they thrive in our dwellings. They can live on as little as one-tenth of an ounce of food a day (something in the neighborhood of a few corn flakes), and they can get water from almost any place where a few drops are available. A house mouse doesn't go far on a daily basis, probably traveling an area ten to thirty feet in di-

ELIMINATING MICE

- Start by setting out snap traps or track cards to see how many you can catch or track on a given night. If there are multiple "visitors" overnight, it's time for serious measures.
- It may be necessary to do some trapping, or you might want to consider the use of toxicants.
- Glue boards are another method of trapping and killing mice. (You may have to do the killing yourself with these devices.)
- Live traps and multiple traps are another alternative.
- Poisons are a drastic measure, but they are successful. The problem, as we'll see, is that it's not always the "cleanest" method of eliminating pests.

ameter. They're true creatures of habit and tend to use the same walkways and food sources daily. That will give you a small advantage in tackling the rodents.

As with many other problem creatures, you'll have to make at least a three-stage plan to keep them under control. If you've already determined that you have a house mouse problem based on the warning signs, you'll need to start by reducing and eventually eliminating the intruders.

Once you've removed the offenders, or if you want to keep your home mouse-free, you'll need to work on sanitation measures. This doesn't mean that your house is dirty; it simply means that there are enough opportunities for a mouse population to find your home attractive. It's entirely possible that you are simply living in an area that is prone to mice infestations. Having an apartment over a restaurant, for example, is a surefire hit for these little rodents. You might want to consider fumigation once you've completed a thorough cleaning of your building.

Mouse-proofing your home is the final stage if you've just rid yourself of the problem. If you are in the preventative mode, it's obviously the first stage. The trick is to "build them out" as you construct—or retrofit—your home. Plug up all openings with secure building materials. Remember that plastic, latex, rubber, and even wood are not mouse-proof. You're going to have to use concrete or metal building materials. If you're sealing up holes, for example, a good filler fiber is steel wool. Cracks around vents or water pipes are potential entry points. You'll also need to examine all doors, windows, and screens to see that they fit snugly. (They'll work best to counter mice if they are metal-edged.) And be sure to check all building foundations; those are the most obvious, and easy, points of entry for mice.

Cleaning up their food supply, of course, is part of the three-step program. Leftover human and pet foods are two of the culprits. And any kind of grain—including stored birdseed—is part of your welcome mat for mice. Stored cardboard, paper, or similar packing supplies are great nest-building materials. Get rid of them. In the yard, keep the grass low and the vegetation away from the house. Any pools of standing water in or near your house should also be high on the removal list.

These are a few of the more basic suggestions about the house mouse. We'll look at some other ways to combat mice when we turn our attention to their bigger relatives—the rats.

RATS

Unlike mice, rats don't feature in our cartoons or star as our childhood heroes. They are definitely considered lowlifes. Have you ever seen a movie about a good rat? The kangaroo rat may be a bit cute (and it is an endangered species), but the Norway rat, roof rat, and the wood rat (the infamous pack rats) are loathed and despised. Gangster films often used the term "low-down dirty rat" for a good reason.

Norway rats are the ones we universally hate. When somebody says they saw a rat as big as a cat, they are probably talking about (or exaggerating a bit about) a Norway rat (*Rattus norvegicus*). They can weigh up to a pound and have a coarse brownish fur and tails shorter than their bodies. And they're found all over North America. Introduced accidentally around 1775, the Norway rat is also known as the brown, house, barn, sewer, gray, or wharf rat. Attracted to wharves by the available food, these rats had no problem getting access to ships via ropes and then lounging about during the cross-Atlantic journey.

Here's how to recognize one of these rodents. They can be as large as thirteen to eighteen inches, they weigh about ten to twelve ounces, and the famous tail can be as long as six to nine inches. Although they're generally thought to be black, they are really more brownish in color, with some black bits on the upper parts. The fur on the belly is much lighter, often beige, gray, or yellowish. And that long whippish tale is scaly with no fur. Like their mouse cousins, their eyesight is poor. But their hearing, smell, touch, and taste are acute. They are good runners and climbers, and they can swim with ease.

Rats are also prolific breeders. They build nests below ground or at ground level and line them with shredded paper, cloth, or other similar material. The female may have four to six litters a year, with as many as six to twelve young in each litter. (The young rats, in turn, are ready to breed in only a few months.) Although rats are born naked and helpless, they mature quickly, becoming independent in three or four weeks and breeding after about three months. Breeding peaks in spring and fall but lags off in the hot summer and often stops during cold winters, depending on where the rat lives. They tend to live a little longer than mice, often twelve to eighteen months.

Not to be confused with pack rats, Norway rats do often live in rat packs of sixty or more animals and are usually descendants of a single pair of animals. The packs may hunt together, and females raise any orphaned rats in the group.

Not much good comes from a rat—unless you're studying behavior in a laboratory. They can burrow under buildings and damage the structure, gnaw electric wires or water pipes, or chew through doors, windowsills, walls, ceilings, and floors to get into a house. They'll eat and contaminate stored grains, any kind of seeds, and even pet food. On farms they're known to kill poultry. Norway rats also carry leptospirosis, trichinosis, salmonellosis, and rat-bite fever. The plague, however, is more commonly associated with the roof rat, not the Norway rat.

If you don't actually see a rat or mouse, you'll still know they're around. They leave droppings along their runways and in feeding areas; these feces are between a quarter of an inch long (mice) to nearly three-quarters of an inch (rats). Other telltale signs include oily smudges along their habitual pathways, and gnawed wood, food packages, windows, or insulation. You might even be able to hear them gnawing or clawing inside the walls.

If you see lots of fresh droppings, it's a good indication of lots of fresh rodents. And they're not very careful in their bathroom habits either. Also, if you see rats during the day, it's certainly a sign of an infestation (although seeing mice during the day is not necessarily a similar sign). A study at the University of Nebraska found that when sixty-five rats were placed in a sheltered animal enclosure, only one rat was sighted every two-and-a-half hours. They're pretty secretive.

Norway rats do well living around us in dumps, livestock buildings, grain silos, basements, sewers, and occasionally in our homes and backyards. They eat just about anything, and plenty of it. Unlike some humans, they select a nutritionally balanced diet, fresh and wholesome, when they can get it. These rats only need about an ounce of water a day, unless they're eating moist foods, such as household garbage. In that case, they hardly need any water.

Rats have such a keen sense of taste that they often can recognize toxins in their food. Although rats eat about a half pound of food a day, and probably waste ten times that much, they may be hard to poison. In an average day, rats travel an area about a hundred or a hundred and fifty feet in diameter. They seldom venture any farther than three hundred feet to get food or water. That means they shouldn't be hard to locate and exterminate.

Rats are intelligent and adaptive; two reasons they're so popular with behavioral scientists. They'll avoid new objects placed in their environs, so traps often don't work for a few days. They memorize all the characteristics of their neighborhoods; you have to be pretty tricky to get a rat. They quickly learn what foods make them sick and avoid eating them again.

Rattus rattus, called the roof rat or black rat, lives on either coast and throughout the Southeast. They're also known as ship rats and first arrived on this continent during the 1500s. They're somewhat smaller than Norway rats, weighing about half as much, with a naked tail longer than their body. Their general biology is similar to the Norway rat's. As do other members of the species, they may horde or cache food, although they may never get around to eating it. They're more adventurous than the Norway rat and will often travel far in search of food, drink, or a home. They'll climb on utility lines, using their long tails for balance, and will live in trees or high in buildings. (Although the Norway rat can climb, it tends to live at ground level or lower.)

Sometimes attempts to cull the Norway rat population will open up habitat for the roof rat. The more aggressive Norway rats will also force roof rats out of the area or at least to different niches in the habitat. That's why you'll see infestations of Norway rats in lower levels of a building with roof rats in the upper.

Black rat (*Rattus rattus*)

The Rodents and Diggers

The third pesky rat is the wood rat, a resident of the East Coast, Great Plains, the Northwest, and Southwest. They tend to live in rural areas. Their most notable behavior is a tendency to snag small objects, such as your jewelry, buttons, cooking and eating utensils, and other small shiny objects. If you uncover their nest, you'll find a cache of these little stolen items. When it's not collecting stuff, the wood rat will be debarking trees, and shredding wiring and your upholstery if they get in the house. Look for their runways or burrows along fences, next to building foundations, or below low-lying vegetation or yard debris.

You might think that the worst part of a rodent problem is the overwhelming inconvenience. Unfortunately, things can get worse. Rodents not only cause tremendous damage to your property and your peace of mind, but they also pose a serious health risk. Over the years, mice and rats have proven to be successful vehicles of a host of diseases, including the deadly Hantavirus pulmonary syndrome (HPS), the plague, and histoplasmosis.

HPS is one of the most recently recognized rodent-related diseases to surface in North America. Infection occurs when a person comes in contact with droppings from a rodent—the most common carrier being the deer mouse—that has the Hantaviruses. It's a potentially deadly disease, so you'll need immediate treatment and care if you see any of the symptoms. Some early signs of the illness are fever and muscle aches that appear one to five weeks after infection. This is followed by coughing and shortness of breath—and hospitalization.

Deer mice, cotton rats, and rice rats, all three of which are found in the Southeast, and white-footed mice in the Northeast, are the most common rodents that might be infected with the specific type of Hantavirus that causes outbreaks of HPS in North America. Although the common house mouse does not carry Hantavirus, any of the rodents that do are capable of finding their way into our homes. The virus exits the system of its host through its bodily fluids, including its droppings, urine, and saliva.

According to the National Center for Infectious Diseases, you're most likely to contract it by touching something that has been in contact with infected urine, feces, or saliva and then passing it on to your nose or mouth. Eating apples from the ground where deer may have fed, for example, would be a problem. (Cider is sometimes made from windfall apples, so that's definitely something to consider if you are living in a rural area and want to make your own juice.) The good news is that the kinds of Hantaviruses that cause HPS cannot be transmitted from person to person.

While you cannot catch the disease from your pet dog or cat, they can bring a diseased mouse or rat to your doorstep. If your pet brings you such a "gift," you should follow this protocol for disposing of it.

GETTING RID OF RATS

- After you put on rubber gloves, spray the dead animal and the surrounding area thoroughly with disinfectant.
- Take a resealable plastic bag large enough to accommodate the rodent and fill it about two-thirds full with disinfectant.
- Put the rodent inside the bag and seal it. (Since you can never be too careful when it comes to handling diseased animals, put it in another bag and seal it, too.)
- Follow the same guidelines for disposing of any materials the dead animal might have touched.
- Your options are then to burn the bags outside, bury them in a four-foot-deep hole, or call your local health department and ask that they dispose of it—or give you further advice for its disposal.

There are some other precautions you should take to avoid rodent-borne diseases. For example, clean a rodent-infested area by spraying disinfectant over every surface. Try to decrease the amount of dust floating around by mopping the floor first with a mixture of soap and disinfectant. To further prevent any chance of infection, wash everything down with a mixture of soap, detergent, and disinfectant. Then let everything dry—and wash it all again with disinfectant only.

It might also be a good idea to have your carpets and/or furniture professionally cleaned. Launder your bedding or other fabrics in hot water with plenty of detergent, and then dry them on a high setting.

In your kitchen, cook all of your meats thoroughly, especially pork products. Rats and mice often live near swine, usually in the large grain troughs. Pigs sometimes eat the rodents' droppings or even the rodents themselves, thereby ingesting *Trichinella spiralis*, the tiny worm that causes trichinosis in humans. The pigs become infected with the disease, are slaughtered, and then the meat is passed on to consumers. Proper cooking ensures the killing of all dangerous parasites, so prepare any pork dishes thoroughly and carefully, and freeze the leftovers right away.

Meanwhile, there are still reported cases of the plague every year in North America. Rats and rat fleas are the culprits. According to the Division of Vector-Borne Infectious Diseases, about 14 percent of cases of the plague in the United States end in death. Fortunately, the disease only afflicts about ten to fifteen people annually in North America, usually in areas and cities where undomesticated rodents infected with the plague are running wild and free. Southern Colorado, northern Arizona, northern New Mexico and California, southern Oregon, and western Nevada have had reported cases in recent years.

Histoplasmosis is less serious than the bubonic plague and HPS, but it still has the potential to cause serious problems in the nervous and respiratory systems. The histoplasmosis fungus is carried in rodent feces. Common symptoms of the disease are lingering coldlike symptoms, achiness, and a feeling of general malaise.

The elderly, the disabled, and young children are the most common victims of actual rodent bites. In addition to keeping their kitchens and eating areas clean, caregivers and parents should wash bedding and clothes frequently. Most rat attacks on infants, for example, occur because the rats smell milk on the baby and the crib.

While you're coming to terms with these unpleasant possibilities, you might already be thinking of ways to get rid of these critters. Unfortunately, that's a challenging task. Keep in mind some of these rodent Olympic feats. Rats and mice can:

- Run along or climb electrical wires, ropes, cables, vines, shrubs, and trees.
- Climb almost any rough vertical surface, including wood, brick, concrete, screen, and weathered sheet metal.
- Crawl horizontally along pipes, conduits, or conveyors.
- Gnaw through a variety of materials, including lead and aluminum sheeting, wood, rubber, vinyl, and even concrete block.

Rats, specifically, can also do any of the following:

- Climb the outside of vertical pipes and conduits up to three inches in diameter, and shimmy up the outside of large pipes by bracing themselves against an adjoining wall.
- Climb the inside of vertical pipes between one-and-a-half and four inches wide.
- Jump thirty-six inches vertically and forty-eight inches horizontally.
- Gnaw and squeeze through an opening a half-inch in diameter.
- Drop up to fifty feet without being seriously injured (that's impressive).
- Burrow straight down into the ground for at least thirty-six inches.
- Reach at least eighteen inches along vertical walls.
- Swim about a half mile in open water, through traps in plumbing, and against the current in sewer lines.

Meanwhile, their smaller relative, the mouse, can do any of the following:

- Jump eighteen inches straight up.
- Travel hanging upside down from screen wire.
- Gnaw and squeeze through tiny quarter-inch openings in almost anything.

Before you begin your assault on these pests, try to find all of the places the animals might use to enter your house. Becca Schad, owner of Wildlife Matters, an integrated pest management firm in Virginia, says, "You really have to go over the house with a fine-toothed comb." Start by checking around the foundation for cracks or openings, and block them with cement or masonry grout. Also, unprotected openings such as dryer vents are fair game to a rodent. If openings, even tiny ones, aren't closed, the rodents may be using them as entry points, and from there, climb inside the walls to get access to virtually any room in the house. Schad confirms that mice can, indeed, get through cracks as small as a quarter-inch wide. And rats don't need much more space than that, either.

Be sure to close off access around pipes and wires with cement, mortar, masonry, or metal collars. Metal baffles similar to the ones used to deter squirrels from getting to bird feeders will work along wires and pipes. And watch for signs of fraying and short circuits with any electrical wires. These can be a potential fire hazard, too.

Another easy mouse entrance is along the bottom of metal or vinyl siding. Mice will crawl up under the unprotected edges and into the walls. Use metal or mortar to block up the ends. They'll gnaw through rubber or vinyl weather strips.

Mice and rats may also enter in a very mundane fashion—through the door. These should fit snugly with a distance between the door and the threshold less than a quarter inch. It's easier to build up the threshold than to modify the door. If they're gnawing along the door, try installing flashing or a metal channel on the lower edge of wooden doors. Be sure it wraps around the sides a bit, too.

Rats are even sneakier. Cover any floor drains with metal grates that are firmly attached. Make sure the openings are a quarter inch or smaller. Also, it's a good idea to cover all pipes that leave your house, such as the ones on your roof, with wire mesh. Many contractors fail to do this. You shouldn't. If people would take these preventative measures, says Schad, "They would save themselves a lot of headaches."

Often attics aren't airtight. Sometimes mice will climb up an outside wall, squeeze through a crack in the eaves, and set up housekeeping in the attic. Cover all vents with quarter-inch hardware cloth. Sealing the entrance holes in your home is a necessary step, but it's a difficult one. Fill up all the cracks. To block up holes, first stuff the opening with steel or copper wool, then cover it with sheet metal. (This double line of defense reduces the chances that mice or rats will chew their way back into the house.)

Rats and mice have a tough time gnawing into flat, hard surfaces, but they can bite into a rough surface with their paired incisors. That's why some materials aren't tough enough to do the job. The U.S. Department of Agriculture recommends the following materials for rodent proofing:

- Concrete, at least two inches thick if reinforced, three-and-three-quarters thick if not reinforced.
- Galvanized sheet metal, 24-gauge or heavier.
- Perforated sheet metal grills, 24-gauge or heavier.
- Hardware cloth, 19-gauge half-inch mesh to exclude rats; 24-gauge quarter-inch mesh for mice.
- Brick, nearly four inches thick with joints filled with mortar.
- Aluminum, 22-gauge for frames and flashing, 20-gauge for kick plates, 18-gauge for guards.

Once you prevent new animals from entering your house, you can concentrate on the home-dwellers. First, there's the old trusty method of getting a cat. It really works, and if you've ever considered getting a cat for any reason, this is one of the best. You might start by keeping its food ration modest, so it doesn't get fat and lazy.

Second, and slightly more aggressive, is to employ a Tokay gecko. The Tokay gecko is from Southeast Asia, and it has a voracious appetite for mice. You can buy them at most pet stores, and they're fairly inexpensive. (As an added benefit, these geckos also like large insects, including roaches.)

Tokay geckos grow to a little more than a foot in length and have blue-green skin dotted with orange spots. They're nocturnal (as are mice most of the time), and they have no trouble walking on walls and ceilings. So even the most agile mouse won't escape them. These geckos do have a few drawbacks, but in most pest-control endeavors, there's a tradeoff. Tokays walk around with their mouths open, and look rather fierce; they also bite when handled. Best to let them roam on their own. Unless you have tons of mice, you'll have to supplement their food, too. They're sometimes not very hardy in northern climates and need a warm and humid home.

Then there are also traps. First, decide if your goal is to kill the animals or to capture and release them. If you want to capture them, make sure you release them far from your home. They have a remarkable way of finding their way back to where they started. Release them in a field where they'll find new quarters, far from yours.

The small Havahart mousetrap captures a single mouse, which can mean a lot of captures and releases. The rectangular box closes as the mouse enters to get the bait and then remains closed until you release the mouse. A large-capacity mousetrap, such as the Ketch-All or the Victor Tin Cat, does not have to be baited because the one-way treadles attract curious mice and then don't let them out of the trap. It probably would be better to bait the trap, though, to increase its effectiveness. The Tin Cat collects up to a dozen mice at a time. Check these traps often so that the mice don't die of starvation or exposure. There are also live-catch traps for rats. Havahart makes one, as does National Tomahawk.

Glue board traps also work, although they're considered a little better for mice than for rats. Glue boards are like fly paper for rodents. When the traps are placed in runways, the animals get stuck to them and can't escape. Be sure to place them on beams and rafters to catch roof rats, and don't forget to treat the attic. As you would for the live-catch traps, check the glue boards often. Kill the animals quickly and mercifully: Drown them in a bucket of water holding them down with a stick, or use a quick blow to the base of the skull. If you have second thoughts about killing the animals after you've trapped one, you can loosen the glue with cooking oil. Glue boards are less effective in dusty areas or in extreme cold or heat. You can buy the traps ready-made or you can make your own with a slow-drying glue and some boards.

Now to the familiar: wooden snap traps. Nearly everyone knows these staples of cartoon slapstick humor. The theory is simple. The animal takes the bait, sets off the trigger, and gets squished by the metal bar that snaps down. In practice, it's not so simple. Rats, which are naturally fearful and smart, may avoid the traps in their environment, and even mice may be wary. To counter this rodent reluctance, first put out baited but unset traps so the rats or mice can get used to taking the bait from the source. (This sounds as if you're feeding them, but you're really just lulling them into complacency.)

Place traps abutting the walls along their runways or travel routes with the trigger next to the wall. If you place double traps parallel to the wall, make sure the triggers face outward. For mice, set the traps close to their areas of activity, spaced no more than six feet apart. For rats, place traps close to the wall, behind objects, and in dark corners. If you enlarge the trigger (the little device that holds the bait) with a square of cardboard, metal, or wire mesh placed just under the wire deadfall, the trap will be more effective.

Baits for rats include a small piece of hot dog, bacon, a nut, peanut butter, or a marshmallow. Mice baits could be a nut, chocolate candy, dried fruit, bacon, peanut butter, marshmallow, or even a small cotton ball, because mice are always on the lookout for nesting materials. For both mice and rats, the bait should be kept fresh and appealing because they won't be attracted to stale baits. Rats and mice will learn how to avoid these traps, therefore a successful first strike, so set plenty of traps.

Some experts in the business have found that the best kind of bait is one that can't be easily removed—or "sprung"—from the trap. Clever rodents will learn how to spring the mechanism and take the bait without harm. A substance such as peanut butter forces them to spend more time at the trap, trying to eat or release the food. This increases the chances that the mechanism will be tripped and the rodent will be trapped.

The Rodents and Diggers

Be careful not to put too much bait in your snap traps. The plan is to lure the rodents into the device and kill them; you don't want to feed them. It's actually the scent of the food as much as the bait itself that is doing the work. Also, try not to touch the trap or the area around it any more than is necessary (you might even want to use gloves for this work). You want the rodent to smell only the scent of the bait, not of the human who set it out for them.

Traps are effective—if labor-intensive—tools against rodents. You don't have to worry too much about being bitten by a trapped mouse or rat, as you would a trapped skunk; mice and rats are much easier to control. You can pick up a trapped mouse with a shovel or a piece of cardboard. Once caught, dispose of any rodent quickly to prevent parasites from spreading. Never touch a dead mouse or rat with your bare hands; always wear disposable gloves.

Sometimes you're going to find that the little critter is not dead in the snap trap, just injured. That poses a little dilemma. One option is to put on your gloves, carefully pick up the trap and the victim, and place the whole mess in a glass jar with an airtight lid. You can either leave the rodent to suffocate (which may take overnight), or you can fill it with water and drown the critter. Neither option is pleasant, but neither is a rodent-infested house.

If you want to drive away rodents, mothballs may work in an enclosed area. Just remember that the fumes may bother humans, too. If you know where the rodents are nesting, you can apply some of the sticky Tanglefoot glues used to discourage roosting birds, and the rodents may be driven away. You may be able to frighten away the more timid ones once or twice with a loud noise, but this method won't work for long.

Another strategy is to install an ultrasonic device. There are many on the market, some of them designed specifically for rodents. As we noted in the insect sections, these devices get mixed reviews. Some experts claim they work under the right circumstances. Lowell Robertson, president of Sonic Technology, is obviously a big fan of his company's Pest Chaser ultrasonic devices. "Ninety percent of the ultrasonic devices sold are ours," he noted. He also points out that it wasn't until the late 1970s that chip technology developed to the point where it was possible to make high-quality ultrasonic devices at low prices.

Rodents are primary targets for ultrasonic devices. "All the rodent family, common rats found in the United States and really all over the world, have the ability to perceive ultrasound," Robertson says. Ultrasound devices are based on the theory that different creatures hear different sound frequencies. Sound covers an extremely broad spectrum, going all the way from subsonic sound to the auditory range, then on to ultrasonic and then microwave frequencies. Human beings can hear sounds of up to about twenty thousand cycles a second; that's as fast as our eardrums will vibrate. Meanwhile, rodents can hear sounds of up to one hundred thousand cycles per second. Within that range there are two different specific frequency bands that rodents hear best. One is just above the human range at 21 to 23 kilohertz. Nevertheless, people complain about being able to sense the sound at those frequencies; these sounds give people a headache. It's even worse for dogs and cats. Those frequencies "would blow people's dogs and cats out of the room," says Robertson. As a result, certain frequencies aren't used for ultrasound devices.

The other range that rodents hear well is right around 46.5 kilohertz. Robertson's device broadcasts in a range from 32 to 62 kilohertz, one thousand cycles per second. This is an optimum range to use. "We can spike it so it hits right on what rodents hear well, yet it's beyond what dogs and cats can hear, and above the ranges that trigger TV sets, garage-door openers, and ultrasonic burglar devices," he says.

"All we do is create acoustic stress," he explains. "You know the big speakers that are right in front of the stage at a rock concert? That's 110 dB by law; the maximum that they can put out. The decibel scale is like the Richter scale, it's logarithmic. A 10 dB increase is a tenfold increase in sound pressure level. So 120 dB is ten times louder than 110 dB. And 140 dB is a thousand times louder. Our device puts out 140 dB. Could you stand being in a room with one of those speakers that was one thousand times louder than that rock concert speaker?" Neither can rodents, Robertson posits. "We create an acoustically hostile environment. We're changing the environment by adding an aspect that those rodents cannot and will not deal with."

To make things more unpleasant for the rodents (and to make them even more effective), these sonic devices no longer just generate a single tone. Many of those that were created early on and failed in the marketplace are what are known as single-tone generators. After a while the rodents would simply get used to the sound and live with it. And after a couple of generations, they actually produced offspring that were deaf to that tone range.

These days, Robertson's devices create a complex sound, one "that's going through 32 to 62 kilohertz in a wide frequency band, and we pulse that sound at sixty times a second in a sinusoidal wave pattern, which is a nonrepetitive pattern so it's changing all of the time." Rodents don't get used to that, he says. "That's what it takes to make a good ultrasonic device: loud noise, complex sound, constantly on."

A word of caution about these devices should be added here. Consumers will probably need to put a device in each room where there is a problem, because the sound will not pass through any hard surfaces. "In fact a sheet of paper will stop ultrasound," Robertson points out. "Ultrasound is very fragile, very directional. It goes in a straight line from the broadcast point. If it hits a hard surface, it will bounce back, and you'll get a carom effect, like on a billiard table. So you can put one high-output device in a pretty good-sized room three to five hundred square feet, and the bounce around will fill in the whole room and it will give you protection in that room."

He points out that the sound won't go beyond that room, however. "That's one of the major mistakes in claims that a lot of ultrasound companies made, and you'll see it in some of the catalogs. They say it covers thirty-five hundred or twenty-five hundred square feet. That leads the consumer to believe that one device will cover a twenty-two-hundred-square-foot house. Basically you need a device in each room where you have a potential problem—where food is prepared, consumed, stored—kitchens, family rooms, basements, garages."

Another method is poison, and at one time or another you might be tempted to use it. Don't. There's always the possibility that another animal will eat the stuff. And although it's rare, children are still killed by poisons left for rodents. More likely, however, is the possibility that the mice or rats will eat the poison and die in your walls—where you will never be able to get them, even though you know they're there. First you'll hear frantic scratching from within. Later you'll smell the stench.

Poisons are difficult to use, too. Rats may not eat them, or they may eat just enough to get sick and then learn to avoid the poison. And rodents, like insects, gradually become immune to the chemicals designed to kill them. Rodenticides really aren't the way to go for household pest chasers.

Whatever other methods you use to rid your house of rodents, the most effective one is to reduce their food sources. This means regular vacuuming, mopping, and sweeping. It also means sealing food in jars and other airtight containers. Metal and glass are the only materials that really count, although tough plastics are deterrents, too. Cereals in their original boxes, for example, are unprotected. Rodents smell the cereal no matter how tightly you curl the wax paper inside. Hungry mice—or rats—will gnaw through the box and then use the cardboard for nesting material. And as we noted earlier for insects, don't leave pet food out for longer than necessary. Rodents consider themselves pets, too.

Meanwhile, out in the yard, you may be putting out the welcome mat once again. Discourage those tendencies by making the yard abutting your house inhospitable to mice. Keep the grass mowed, engage a seed catcher beneath your bird feeder to prevent seed from reaching the ground, and clean up pet food. Raise your woodpile by fifteen inches, and then mice won't burrow under it. Keep those field mice in the fields where they belong.

Clean up and cover any litter. High grass and heavy brush attracts a variety of animals. Keep bushes and trees trimmed so they don't touch the house. If you can, cut brush three feet back from the side of the house. Heavy vines and vine ground cover are also good cover for rats. Rats also need more water than house mice; so get rid of standing water, too.

Metal trash cans are better than vinyl or plastic ones, which the rodents can gnaw through, and it's best to have the cans in a stand so they don't tip, spilling tempting garbage. Of course, don't feed your pets outside unless you can clean up immediately. And in the house, don't leave uneaten food in the pet's bowl. The dog or cat may complain for a while, but it's best to do away with the attractant—at least until the rodent problem is under control. Make it up to your pets by showering them with extra affection.

4

THE BACKYARD FEEDERS

As the line blurs between rural and urban areas, it gets tough to tell where the "wild" areas are anymore. We've moved into the territory of the small mammals, and apparently plenty of them have decided that the city pickings are just as good as those in the country. Some of the critters in this section—such as squirrels and chipmunks—could have been included with the rodents in the previous chapter, but they seem to fit better with the other backyard feeders. Once again, we have persistent animals to deal with here. As some readers will attest, they can seriously test your patience and make your yard a true battlefield.

RACCOONS

Practically anyone who has gone camping has come back with an, "Aren't they clever little !@#$!" story about raccoons in the wild. You might have sealed up your perishables in a plastic bag, for example, and submerged them in a river to keep them cool overnight. In the morning, you found that raccoons had lifted the bag out of the water, untied the knot, removed and opened the carton of eggs, neatly broken each one, and sucked out the contents.

That kind of cleverness is almost enough to make you forgive them. But raccoon antics in suburbs and cities, where they're increasingly making their homes, are not so endearing. Larry Manger, wildlife biologist with the U.S. Department of Agriculture's Animal Damage Control Department in California, tells the story of a California man who owned a prized Porsche. Raccoons got into his garage one night, and in the course of some mad rummaging, managed to knock several filing cabinets over onto the car. Manger says the damage to the car cost ten thousand dollars.

Then there's the story of another Californian who lived in a beautiful, spacious house, four or five thousand square feet in size. He spent most of his time at one end of it, so he didn't realize that raccoons had actually torn off a small part of the roof on the other side of the building! Raccoons sometimes do this sort of thing when they have young, Manger says. "If they can't get in through a screen or something, sometimes they'll just get an inkling they want in, and they'll just rip a hole in the roof and go in." Well, one night the owner was watching TV in the raccoon's side of the house. It rained rather heavily, the plaster got soaked, and the whole ceiling came down. "He ended up with close to twenty-five thousand dollars' worth of damage," says Manger.

It's sometimes hard to believe an animal of this diminutive size can cause such massive problems. Raccoons don't wreak as much physical havoc in your garden as gophers or groundhogs, but they are often considered the biggest bother in your yard, and specifically in your garbage cans. The raccoon is a small mammal, typically twenty to thirty inches long, weighing fifteen to thirty-five pounds. (Although in urban areas, where they thrive on our garbage, they can weigh up to sixty pounds.) Their fur is a grayish brown. But what every schoolchild remembers about a raccoon is its bushy banded tail and black mask.

There is an appropriateness to the raccoon's bandit-like appearance. They're smart, curious, bold, fast, and sneaky. And they will snoop into anything that smells like food—no matter how well secured it is. They have the kind of clever "fingers" that some safecrackers would envy. A raccoon's front paws are unique because they are similar to human hands with five dexterous digits. This makes lifting the lid of your trash can effortless. In fact, you can tell if a raccoon has been around, because the prints of their forepaws actually look just like tiny baby hands. And they use these to pull, pry, turn, twist, grab, and rip their way into all sorts of trouble. They also have the persistence and—despite their size—the strength to go after whatever they put their minds to.

While many animals suffer when humans develop the land, raccoons seem to thrive. They live well in woods, suburbs, and cities. Raccoons range from Hudson Bay all the way to South America. In the United States, raccoon populations are estimated to be still at about the same levels today as they were in the mid-1800s. One reason is that they are very adaptable; they can live in swamps, jungles, and hardwood or coniferous forests. One particularly bright group of them living on the San Juan Islands off Washington state have reportedly learned to unzip the storage compartments in nylon hiking packs.

The reason for this success is that raccoons are so adaptable. Take their diet, for instance. They're omnivores, so they eat anything. They like fish, turtles and their eggs, shellfish, salamanders, insects, bird eggs and birdlings (they can decimate a waterfowl nesting site), young mammals, fruit, nuts (especially acorns), grains, almost anything growing in your garden (especially sweet corn), cultivated crops, garbage, and other food associated with humans—pet food, scraps around the barbecue, and birdseed. They'll even brave wasp and hornet nests to get to the

Raccoon (*Procyon lotor*)

goodies inside. They turn to a variety of foods, with a particular fondness for crayfish and frogs. As suitable habitat disappears, they seek out our gardens, yards, and garbage as food sources, but even then they vary their food sources.

They also eat pizza, which may be the real reason they are thriving in suburbia. Witness what happened to one homeowner when he was having dinner outdoors on his patio. The man stepped inside for a few minutes to answer the phone, and when he returned to his meal, "the unwatched pizza was in the middle of the backyard," with a raccoon munching on the mushrooms, pepperoni, and cheese. He even liked the anchovies.

Raccoons are known to dunk their food in water before consuming it, a habit that earned them the name *Procyon lotor* from the early biologists. *Lotor* means "one who washes." The genus name was made up, because no one could figure out to which family the raccoon belonged. (Ancient naturalists even thought it was a bear.) Yet, why do raccoons wash their food? Some folks speculate it has to do with making sure live food is really dead by drowning it first. Maybe it makes the food easier to chew—like dunking a stale donut—except it's turned into a kind of nervous habit. It might just taste better to them. A more plausible reason is that they use the water to separate the shells from the flesh when they are eating invertebrates such as crayfish.

In suburban settings, raccoons have sometimes had to give up this fastidiousness for lack of a ready water source (although when eating pizza, perhaps they dunk it in soda). No one really knows why they dunk their food; it's certainly not to clean it, because the food is usually consumed regardless of its condition. They certainly have definite tastes. Author and naturalist Virginia Holmgren has watched and fed raccoons in her backyard for more than thirty years and found that the animals would filter through bakery scraps to first eat the sweet breads, then eat the rye and pumpernickel last.

Raccoons are also adaptable when it comes to their accommodations. Being good climbers, they often prefer to den in a hollow tree. But they'll also intrude on other animals, using burrows and dens that others have created. In some cases, they may simply use a secure, sheltered rock crevice, tree branch, or squirrel's nest. They also like outbuildings, sheds, garages, and sewers as denning sites.

Because raccoons are nocturnal, you don't always know they're out there unless they do some damage to your property or unless you happen to spy one during the daytime as it naps and suns in a treetop or rock crevice. (You might overhear them chatting as they munch on your garbage. They vocalize with a soft *chrr* when they feed.) Because they rarely stay successive days in the same den, raccoons keep several sleeping dens in their home range, which can be as small as ten acres and as large as several square miles. The size depends on how much food and how many animals there are in any given area.

The Backyard Feeders

Home ranges can overlap, and the raccoons develop a social hierarchy to allow their coexistence. A male controls a larger area than a female, while several females may live in his territory. The first meeting between strangers probably involves a fierce display and posturing, even fighting. This establishes who's boss, and at subsequent meetings, the loser will give way. They may also scent-mark their territory by leaving scat as a marker or by rubbing their anal scent glands on objects.

In the colder North, raccoons den up for winter, but in the South they remain active. Northern raccoons spend the summer eating and putting on weight for winter, when food is scarce. They may double their weight—not too difficult a task considering their catholic tastes. Nevertheless, they don't hibernate, and have to remain somewhat active over the winter. It's curious to note that early in the twentieth century, Russia imported raccoons in an attempt to raise them for fur. Unfortunately, they didn't fare well over the cold winters and many died. Survivors apparently made their way to Germany and survive there still.

Adult raccoons are mostly solitary except in mating season and when they have young with them. A breeding-age female is rarely alone, perhaps from the time the last kit of one litter leaves until a male arrives to mate and produce the next litter. (Adult females may occasionally hunt together, too.) Mating season begins in December in the South and in January or February in the North. The young are then born in March, April, and May.

Females mate when they are a year old; males wait until they are two, living for a few years alone or in bachelor groups. The male is polygamous and will mate and live with a female for several days or weeks before going off in search of a new mate. If the female does not become pregnant after mating, she will become fertile and mate again. Meanwhile, a den for the family has to be really secure, so the female will seek a suitable site, one with a protected opening, preferably high off the ground.

After breeding, the female stays alone. About sixty-five days after conception, she gives birth to three or four kits, helpless, four-inch-long, two- or three-ounce blind animals. They have a little bit of fur, but by the time they are ten days old, they sport the characteristic facial and tail markings. The mother stays with them for a few days more, but then resumes her nightly foraging for short periods, staying close to the den.

After about three weeks, the kits' eyes are open, although they sleep most of the time. By the age of one month, they weigh about two pounds; the mother now leaves them alone all night while she forages. If the den is high up, at this stage the female often moves the offspring to a different den, one that's lower to the ground, because as they begin to explore the kits will sometimes fall out of the nest. While raccoons can survive drops of up to twenty feet, mother raccoons seem to prefer not to put this theory to the test so early in life.

The little raccoons can walk, run a little, and climb when they are two months old. After they are about ten weeks old, they begin weaning and start eating solid food. Pretty soon they roam far and wide with their mother and begin to sleep in different dens in the home range. They are completely weaned when they are twelve to sixteen weeks old, but they still feed in the mother's home range, even if they don't come back to her den every night. In the North, offspring may spend the winter in the same den with the mother until she needs the space for a new family and drives them out. Or, they may just den nearby. In the South, the young seek their own way in the fall, with females generally staying nearby and males seeking new territories.

As long as the temperature stays above twenty-eight degrees Fahrenheit, raccoons will forage, but they sleep the night away if it is colder. As the winter passes, the animals may grow acclimated to the cold and will be out when it is as cold as zero degrees. At any rate, by the springtime, they usually weigh about half what they did going into the winter, and late winter and early spring are especially hard on the animals.

Raccoons born late in the season may not have had time to build up fat stores to survive a winter, and they may starve or become prey for other animals. Adults, however, are fierce opponents and are rarely prey, except to humans and cars. All in all, they probably survive for about six years in the wild.

In suburban areas, towns, and cities, raccoons may live in parks, in woods, near streams, or even in your backyard if it provides a suitable habitat. That would require access to food and water and a place to live—a denning tree, a sewer, a hole underneath your house or other building, an uncapped chimney, or a cozy attic. (Unsuitable habitat, to a raccoon, is a yard containing a dog.) Although it may be exciting, initially, to discover that you have raccoons as neighbors, their penchant for overturning your garbage can, tearing holes in your roof, and terrorizing your birds can quickly temper your enthusiasm.

Rabies is also an issue of growing concern. The eastern mid-Atlantic states have seen a rise in the number of raccoons carrying the virus. It moves through the country like a wave, peaking and receding. For example, in 1981 in Maryland, there were only seven reported cases of rabid raccoons; in 1984, the number had risen to 964. It's even worse in Westchester County, near to New York City, with seventeen reported cases in less than a year. To test an animal for rabies, it first must be killed, and anyone bitten by an animal that can't be tested is treated for rabies with a series of five shots.

For this reason, raccoons and other wild animals are considered dangerous. (Skunks, foxes, groundhogs, and squirrels may carry the virus as well.) Raccoons generally don't attack people. Rabid raccoons, however, are exceptions. A rabid animal may be especially friendly, lethargic, or disoriented, but lethargy is also a sign of distemper, a disease not contagious to humans. Report any oddly behaving animals to your local wildlife authorities.

Raccoons also carry an organism called raccoon roundworm, which can infect people who come into contact with raccoon feces—while walking in the yard barefoot, for example—or who inhale the roundworm eggs, which is unlikely since raccoons (like cats) are private about their toilet habits. Another caution about raccoons: Female raccoons that are not rabid may still attack you if they perceive a threat to their babies. So in the summertime, be particularly careful around them.

If raccoons live in your vicinity, chances are you'll know it because they will have made a raid on your garbage cans. Raccoons love trash and are regular Houdinis at breaking through whatever systems you devise to keep your cans closed. They usually knock over a garbage can to get at your tasty leftovers, so if you can keep your cans upright, you may begin to thwart the masked marauders. To keep them upright, place your cans in a rack or tie them to a support. But it still may take some cleverness to devise a lid they can't conquer.

You may have trouble if you store your birdseed in a container, even a metal one, in your backyard. Although raccoons aren't especially fond of this food, they will be attracted by its smell and they'll take the can lid off and rummage just to see what's inside. Even if you secure the containers with bungee cords, the raccoons may still eat right through them.

Another way to tackle this problem is to make your garbage less appetizing. You can buy repellents such as Ropel, a sticky, foul-tasting concoction that you spray onto the outside surface of your plastic bags. One taste and the raccoons will look for dinner elsewhere. Cal Saulnier, with Plow and Hearth Catalog, can personally vouch for how foul this stuff is. "I've tasted it, and it's disgusting," he says. It is also relatively inexpensive and harmless to animals and the environment.

Still another approach involves making the cans less inviting. Several catalogs sell a small mat that delivers a mild electric shock to animals that step onto the surface. These mats, which go by names such as Scat Mat or Invisible Gate, are made to be placed under the cans. The theory is that the negative stimulus will train an animal to avoid these areas. Of course, you may have trouble getting anyone to take out the garbage. The devices usually sell for about sixty dollars, and if it fails outside you can always use it to train your pets to stay off the couch.

You can also try to make your garbage cans less accessible. Try putting the cans in a place where the raccoons can't get at them, such as in a sealed garage. A related strategy is to put your can inside another larger can—although the critters might simply view this as an added challenge and some diverting entertainment. Plain old bribery may work, too. Some California residents found that by offering them pet food, they were able to persuade the raccoons to leave the garden and lawn alone. Raccoons in the yard take some getting used to; they may still attack the lawn, garbage, and garden—but not with the hunger of before—plus they'll leave footprints and evidence of play all over the yard.

Raccoons like the grub worms that infest thick lawns. In a well-established lawn, they'll dig for the creatures. And, if you've just laid down new sod, watch out. Raccoons may roll back your new sod searching for tasty grubs. To keep the animals off your lawn, there's an old remedy that calls for spraying with a mixture of one cup each of children's shampoo and ammonia. It should keep the raccoons away for a season. That formula has been updated by substituting one cup of Hinder animal repellent for the ammonia. (Hinder is a liquid concentrate that contains the ammonia soaps of fatty acids. It is, however, approved for use on edibles.) You can obtain Hinder through a farm supply store or catalog.

You could also apply milky spore to the lawn. It will parasitize and kill the grubs, so the raccoons won't have any reason to dig up the lawn. Milky spore often takes years to get going, though, so in the meantime you'll also need a short-term solution. Try altering the watering of the lawn. Grubs can't survive in dry soil. If you're trying to save your sod, one solution—assuming you're not dealing with too large an area—is to pin the sod down with stakes or wire pins.

Fruit trees are another raccoon temptation, and excluding them from these forbidden fruits can be a challenge. But if the tree is isolated enough so that the animal can't jump into its fruit-laden branches from a nearby tree, there is still hope. One way to tackle the problem is to "flash" the tree trunk: Put a collar or slippery metal around the trunk, which prevents the animals (squirrels as well as raccoons) from climbing up the tree. Metal flashing is usually thin aluminum, which comes rolled on a spool. Look for flashing that is at least three feet wide; that way, it sticks out far enough from the tree that the animal can't reach out to the edge and pull itself over it. Tack the metal around the tree, being careful not to make it so tight that you girdle the tree.

In the northern part of Sacramento County where Larry Manger works, many people have installed small ponds for koi, an expensive variety of oriental carp. Naturally, wildlife enjoy dining at these pricey piscaries. Says Manger, "I've had people with koi worth thousands of dollars, ones they've imported from overseas; they come out in the morning and find the raccoons have eaten them." (To be fair, fishing birds like herons and kingfishers may also eat fish from backyard ponds.) One way to protect fishponds is to make them more than two and a half feet deep, Manger advises. Raccoons catch backyard fish by wading into the pond, corralling the fish in a corner, and then grabbing them. But in deeper ponds, where raccoons have to swim, they can't do that and the fish can escape by swimming underneath them.

The Backyard Feeders

If you don't want to deepen your pond, one inexpensive and effective strategy is to place some pieces of terra-cotta pipe, about sixteen or eighteen inches long and four inches in diameter, on the bottom of the pond. This gives the fish someplace to go when the raccoon is in there swishing around. The fish can go in the pipe, and when the raccoon sticks its paws in one side, the fish can maneuver and have a better chance of getting away, Manger explains. Also, the pipes won't detract from the aesthetics of your pond. "After they're in the pond for a while, the pipes moss over and you can't really see them," he says. According to Manger, the most effective approach is to install a small electric fence around the pond. He thinks these protect the ponds from raccoons, but he qualifies his approval by pointing out, "an electric fence kind of detracts from the pond."

Another target for raccoons is your birdhouse. Raccoons find bird eggs a delicacy. Here's where technology really helps. The Bird Guardian is an effective ploy for outsmarting egg- and birdling-eating critters. This simple mechanical device adds a three inch-long tunnel to the opening of a birdhouse. A hungry raccoon or cat won't be able to simply reach its paw into the house opening to snatch up helpless nestlings; however, some birds, most notably bluebirds, are frightened by the device and won't enter the birdhouse through it. To address that problem, the manufacturer has recently added a short ladder outside the tunnel, which the bluebirds can use for balance. (While the Guardian will stop a short-armed mammal, it won't halt a snake, so you'll need other measures for snake problems. The eastern black snake, for example, is an adept climber.) Although the Guardian is an effective tool, a determined raccoon (is there any other kind?) may be able to pry it off.

One reason you may be getting so many unwelcome visitors to your garbage, lawn, or birdhouse is that the raccoons may actually be denning in your yard. If that's the case, one way to encourage them to move is to make your yard an uncomfortable place for them to live. And as I mentioned earlier, getting a dog is one approach. Another method is rock and roll. Raccoons, as do many small mammals, seem to have an aversion to loud rock and roll music—heavy metal in particular. An outdoor loudspeaker aimed at their nesting place should get rid of the raccoons that once considered your yard a peaceful haven. The music might attract neighborhood teenagers or angry neighbors, however, so that's a tradeoff you'll have to make.

Humor aside, this is a principle to keep in mind for most critters: Unnatural noises scare animals. The louder the better. The stranger the better. If raccoons are persistent problems, you might also consider the following apparatus: an outdoor speaker connected to an infrared motion detector. When anything larger than a house mouse approaches, a loud noise goes off. Works wonderfully.

A common household problem in big cities is raccoons using your deck—at any level—as a rest and play area. It's a particular problem if trees surround the deck, which makes it easy for raccoons to visit it. Don't be tempted to feed these initially "cute" animals. Once you give them handouts, there is almost nothing you can do to get rid of them.

Here's a fairly simple and ultimately nontoxic solution. Spread mothballs on the deck. Most people don't want to use them indoors, where the harmful levels of naphthalene can accumulate, but outside the fumes will waft harmlessly away. Not only do moths have an aversion to mothballs, but raccoons also aren't crazy about their smell. After a few weeks, most raccoons will find some other place to live. They don't like strong, sharp smells. Dousing a deck with ammonia would have about the same effect.

There are two reasons that raccoons might actually come into your house. Sometimes it's just on a temporary basis; other times they are trying to set up a permanent residence. There are different ways to deal with each kind of intrusion. Some more brazen raccoons march right in through the pet doors, usually because they smell pet food inside and want a snack. Betsy Webb, curator of the urban wildlife exhibit at the Denver Museum of Natural History, had a raccoon that would enter her home nightly through a dog door. While she maintains, "My mission in life is not to keep animals out of my house," this particular mother raccoon repeatedly helped herself to Webb's food.

At first, Webb used the dog door's wooden slats to secure the opening at night, but the raccoon punched the door open. Then, Webb tried propping a chair against the small door, but the raccoon pushed inside for its midnight kitchen raid. Webb finally had to close off the door entirely until the raccoon stopped entering the house and eating. Webb removed the barrier after a few months and had no more problems. She thinks the female raccoon was entering the house during times when she had young and needed the extra food.

If you come across a raccoon in your kitchen, the best tactic is to open the doors and windows, back away, and let the raccoon leave of its own accord. If you try this technique at night, turn off the lights in your house so that you don't invite every moth in the neighborhood inside. Once the raccoon leaves, secure the pet door.

People also occasionally encounter raccoons in other rooms. Marta Vogel, of Takoma Park, Maryland, had raccoons that apparently came in through a heating vent, made their way down to the furnace, and then by "squishing their bodies" actually came out through the furnace and into the basement. Brazen creatures, they didn't stop there. Sometimes they even wandered upstairs. Marta's roommate Valerie occasionally mistook the raccoons for one of Marta's cats "because our cats have striped tails." What did Marta do about the problem? She moved. "I left them for the next tenant."

The Backyard Feeders

If you come across a raccoon inside your house—and don't want to take Vogel's pragmatic approach—remember this fact: "The trapped raccoon is as frightened as you are," says Laurie Bingaman, with the National Zoo. Always try to move slowly and unobtrusively. Isolate the intruder in a room that has access to the outside, then open windows and doors so the animal can escape. Becca Schad, owner of Wildlife Matters, an integrated pest management firm in Virginia, adds, "Pretty much, when left to their own devices, they'll get themselves out. They realize it's not an appropriate spot," and will try to leave.

But if the animal seems too confused or afraid to act in its own interest, you may have to motivate it. Walk slowly toward the critter to herd it outside, but don't wave your arms or yell, which may confuse it. Before you begin the procedure, you may want to slowly and gently remove Grandma's teacups and saucers—or any other valuables or breakables—from the room, since things may get exciting. It's also helpful to have a broom, not to flail around, but to use as an arm extension to block a movement in the wrong direction. As the animal exits, in the words of Bingaman, "Lunge toward the open door or window and close it."

Chimneys are warm—there's always heat escaping from them—and raccoons (not to mention other animals) are attracted to them. So as long as there's no fire, they make great dens and places to raise a family. The chimney is a case where an ounce of prevention is worth a pound of cure. It's easy enough to raccoon-proof your chimney, and it's wise to do it if raccoons are ever seen in your vicinity. If you don't, you may well have problems.

Most hardware stores carry a cap that effectively blocks entry to your chimney. It is made out of wire, and once it is attached to the chimney top, it allows smoke to exit but won't allow animals to enter. Chimneys are different sizes, so before going to the store, you'll need to measure the opening in order to get the right size. Check with your hardware store to see what's available. Strong masonry cloth should work as well; just attach it securely to the chimney with masonry screws.

If the animal (or animals—a mother may give birth and nurse her young in a chimney) is already living in your chimney, the problem is a little more complicated. One way to tell if you have chimney residents is if you hear mice-like squeals coming from the area above your damper. When you suspect you have residents in your chimney, start by blocking off their entrance to your house. If, heaven forbid, the flu is open, block the opening into your room. Otherwise, block the fireplace opening securely, keeping in mind that a mature raccoon may weigh thirty-five pounds or more and can move heavy objects.

Once you've determined that your resident is a solo adult, one option is to wait for the animal to leave on its nightly rounds and then scramble up to your roof to install your new animal-proof chimney cap. If your unwanted tenants include immature offspring, you can wait for them to grow old enough to leave on nightly rounds with their mother and then cap the chimney. The young nurse for two months, so if you take the wait-and-see strategy, you won't be waving good-bye to the family until the young are weaned. If you can hear the babies, they're probably already at least six or eight weeks old, however, points out Becca Schad. Furthermore, that means they've already "been there for several weeks, so what's another couple of weeks?" She recalls one client, a woman over eighty, who complained about noises in her chimney. "When I told her she had a raccoon family in her chimney, she said, 'Well, I guess I'll just have to wait.' I just wanted to hug her!"

If you don't have that kind of patience, you can try to encourage the animals to leave on their own. To do that, use "some mild form of harassment" to disturb the mother and make her aware you're there, points out John Hadidian of the Center for Urban Wildlife of the National Park Service. You can shine a mechanic's trouble light down the chimney and put a radio on in the fireplace; raccoons don't like bright light and loud noise and will generally move. Since they also don't like certain strong smells, you can try placing a cup of ammonia in the fireplace, or dropping some mothballs down the top onto the flue.

You can also call in a professional to remove them. Bear in mind, though, that professionals will usually kill the animals. Raccoons living in your attic (or under your porch) are best dealt with by leaving the lights on and keeping a radio going nearby. This will often persuade the raccoons to leave. But then your task, of course, is to find their point of entry. "There are always structural problems when an animal is in a house: a hole in a board that leads to the attic, a hole in the roof, an unscreened opening underneath a crawl space," says Hadidian.

Even if the animal seems to have left for good, it's important to go ahead with the repairs to prevent the same animals, or other ones (or weather), from getting in. "In ninety percent of the cases where people haven't made the repairs, I get a call the next year because the raccoons are back," Hadidian says.

And what if you can't find that point of entry? Now you get to play detective. Carefully look around your house to find the possible access points; watch the raccoons at dusk to see if or where they are emerging on their nightly rounds. And don't ignore even very small holes. "If an animal can wiggle its head through a hole, it can usually get the rest of its body through," says the National Zoo's Laurie Bingaman.

Once you've zeroed in on the hole, wait until the raccoons have left on their rounds, and then plug it up. If you don't want to wait until after sundown to do your carpentry work, you can try attaching a one-way door at their point of entry.

Hinge the door so that it only swings out, but not in. Then, once the animal has left for good, you can close up the entry point permanently.

If you're about to block up the hole, take along a bright flashlight as a precaution. Not only will it help you as you work, but in the event that you find a raccoon inside the attic or under the porch, the brilliant light will also temporarily blind it. That gives you that second or two you'll need to get the hell out of there. (Raccoons are not likely to charge, but bumping into one in a small, dark place isn't advised.)

It's important to keep in mind that blocking a raccoon's entrance into your house—or to its den in your yard—does not always have magical effects. A raccoon establishes several dens in its home range. Though you may have kicked it out of your house, it will likely continue to live in the area. And it may well show up again in your yard. Sometimes, "it seems like once they've decided they want to make your place home, not a lot works, except maybe a big dog," says Larry Manger.

If you have an animal that stubbornly refuses to leave, or that is causing particular damage, you may want to consider trapping. "Trapping is a last-ditch effort," Manger says. It can solve the problem, though. "Often you have a particular animal with a particular habit that's causing the damage. Lots of times you only have to reroute one animal." For example, he says he has been called in on several cases where raccoons were killing chickens. "I'll take that one raccoon out, and that will be the end of it, even though there are still other ones in the area. The particular one that had the habit is gone."

Wildlife experts recommend not trapping the animal yourself unless you know what you're doing. In some areas trapping may be illegal, and it is always

---

KEEPING RACCOONS AT BAY

- Keep garbage cans stored in an outbuilding or make sure raccoons won't be able to get the lids off.
- Treat garbage bags with a foul-tasting repellent.
- Make your home secure against nesting and denning raccoons.
- Cap your chimneys.
- Don't handle any animals in your house; let them find the door themselves.

---

dangerous because raccoons can give a nasty bite and, as mentioned before, may be carrying rabies. If you have an animal in your house or yard that you suspect is rabid, call animal control experts immediately. They will trap and kill the animal. If a problem raccoon does not appear to have rabies, the wildlife professional may decide to release it after it's been trapped. Be sure the animal will be transported a good distance away, at least twenty miles or more. Otherwise, these nomadic creatures may make their way back to your area.

Here's one final consideration. The mortality rate for raccoons that have been moved is very high. It is estimated that nearly 50 percent die in the first three months after the move. With that statistic in mind, you might want to be sure that you've first tried to get the animals to move away on their own, and consider calling in someone to trap the raccoon only after other methods have failed.

SKUNKS

There is nothing as pungent as the odor of a skunk. Ask the residents of Gilroy, California, which boasts that it is the "Garlic Capital of the World." With its annual garlic festival, Gilroy thought it had devised the smelliest event imaginable—yet skunks managed to upstage those puny bulbs. While feasting on some garbage cans at the festival grounds during the 1990 ceremonies, some skunks got riled and decided to compete for the pungency title. When they added their own perfume to the air, even seasoned garlic sniffers were forced to scatter, and the evening was nearly ruined.

Incidents such as these have given skunks a public relations problem. Year after year, these critters show up on America's least-favorite-animal list. This causes skunks some consternation, since they are basically timid animals that do not go looking for trouble. And, unlike some other mammals, they don't run around displaying large teeth, sharp claws, or unusual strength. In fact, ranging in weight from six to fourteen pounds and measuring from about thirteen to eighteen inches (not including their bushy tails), you might even say that skunks are rather puny.

No, their public relations problem basically comes down to one of odor, and a decidedly unsavory odor. Like so many difficulties that arise in both the human and animal world, these troubles can be traced back to a pair of glands. In the case of skunks, these are scent glands, located on either side of the anus. They secrete a wickedly disgusting liquid, a sulfur-alcohol compound, which the skunk sprays on perceived attackers.

Interestingly, skunks prefer to flee and save the chemical onslaught only for threats they can't elude. If threatened or attacked, a skunk will begin a kind of warning "dance." It will march toward you with its tail erect. That's the first sign to back off. Then it will click its teeth and hiss. Finally, it will start stamping its front legs or shuffling backwards.

But it's probably too late when the skunk forms its body into a U-shape, keeping all four feet on the ground, with its face and upright tail facing the target. The oily, yellowish liquid shoots fifteen feet from one or both of the scent glands. At six feet, it's a pretty accurate stream; at longer ranges it's more of a spray or mist. One variety of skunk, the spotted skunk, even sprays over its head while balanced on its front legs.

This proves to be quite effective, because the skunk is rarely anyone's dinner. The great horned owl is the skunk's fiercest predator because it has a terrific sense of hearing and sight, but a lousy sense of smell. Other possible attackers might include hungry bobcats, fishers, foxes, or coyotes. Once animals tangle with a skunk, however, they generally don't forget the lesson. The exception, of course, is your dog. But more on that later.

One interesting note is the skunk's color, or lack thereof. Many of the creatures that might otherwise attack a skunk are actually color-blind, and since skunks also

Common striped skunk (*Mephitis mephitis*)

travel at night, they're more difficult to see. So the genetic evolution of the skunk has brought us to the current variety, which is a highly distinctive black and white.

The skunk ranges through the United States and Canada, with the exception of dry, arid regions. It lives in many different environments, from prairies and woods, to farms, suburbs, and even cities. The more familiar and more common striped skunk (*Mephitis mephitis*, which means "bad odor") is found in most areas, except in the range of the spotted skunk (*Spilogale putorious*), which is found in the Delaware-Maryland-Virginia tristate area and from New England west through New York and Wisconsin and south to Illinois. The skunk family also includes two other members: the hooded skunk (*Mephitis macroura*) and the hog-nosed skunk (*Conepatus mesoleucus*), found in the extreme southwestern parts of the Great Plains. Many of these skunks' behaviors are similar.

Skunks are solitary creatures, and other than a mother and her offspring, you will rarely see two skunks together. They are usually nocturnal and search for food at night, traveling a mile or two, though they never stray far from water. They are opportunistic feeders and eat a variety of plants, animals, and insects they encounter. Of course, during the summer that means they eat a lot of insects—grasshoppers, beetles, crickets, and grubs. They'll also eat mice and even larger mammals such as rats and rabbits if they can get them.

On the ecology scoreboard, skunks are clearly in the plus column. This is because they have a strong appetite for insects as well as mice. Their diet also includes shrews, rabbits, bird and turtle eggs, berries, grapes, and, of course, garbage. Sometimes they also raid chicken houses and feast on eggs, much to the chagrin of farmers. In warm climates or seasons, skunks spend their days resting in grassy fields, brush, or other places that afford good cover. In colder weather, they seek out a good spot to keep warm—often a burrow or hollow log.

Skunks like to do as little work as possible on their winter dens. Although they may dig their own holes, usually they'll use an existing woodchuck hole or perhaps a crawl space under a house. It's not unusual for the solitary skunk to gather with other skunks in a winter den; usually a male will share a den with one or more females. Once in a while, they'll even share a burrow with another species, probably the one that did the work. But they don't bother the animals, and they're usually in different parts of the tunnel. The skunks' improvements include a little decorating with shredded leaves or grass, and if it's really cold, they may even plug the entrance.

During cold weather, they stay in these warm spots, slow their metabolism, and go into a deep sleep, but not a true hibernation. In preparation for this rest, they eat voraciously and fatten up. During warm spells in the colder months, they may then come out of their dens and forage.

The Backyard Feeders

In later winter, males venture forth seeking a mate. (Skunks are ready to mate when they're about a year old.) If a male skunk denned with one or several females, he may not need to look very far. A pair may breed and stay together until the female is ready to give birth, or the male may move on immediately. At any rate, he doesn't stay around to help raise the kids.

The female gives birth to five or six tiny babies after about two months. The half-ounce kits are blind and deaf at birth, but already have the characteristic black-and-white skunk marking on their pink skin. At about two weeks, they have grown a nice fur coat, and when they are three weeks old, they can hear and see. They weigh twelve ounces after a month of nursing, but they can barely walk and don't leave the den.

When they're about seven weeks old, and weigh about a pound and a half, they venture outside with their mother. Skunks are thorough in their explorations for food, and they inspect any possible source. The young learn to sniff every hole and to look under every rock in the small home range. They are weaned soon after they begin hunting with their mother, but they stay with her afterward throughout the summer. In the colder northern regions, the young skunks may den with the mother for the winter, but in the South, they part ways in the early fall.

Skunks usually live two or three years in the wild, but they may live as many as five or six. In most areas, skunks don't have any natural enemies other than dogs. And, as fur-bearing animals, skunks are legally protected from humans, outside of the trapping season. In fact, deodorized skunks are coveted as pets. They're playful, intelligent (skunks can learn to recognize their names), and they're amusing. Many localities, however, outlaw pet skunks. Besides, unless you can convince your neighbors that the skunk is 100 percent aroma-free, you probably aren't going to be too popular. As a result, most people encounter skunks in the outdoors, including, increasingly, around their yards and homes.

Skunks adapt to changes in their environment, including residential development. When their natural environment is invaded, they move into residential neighborhoods in search of food and shelter. For example, in Reading, Pennsylvania, thirty-three skunks were found living in a single city block. In Chicago, a family found a skunk in a heating duct in the house. In Nelson, British Columbia, skunks had taken up residence under the plywood floor of a school and wouldn't allow music students to practice. Every time one of the students tuned up, the skunks showed their displeasure. One young flautist complained that it was "really hard to concentrate, especially when you have to take a big breath." In California, meanwhile, a homeowner found fourteen laid-back skunks living (not breeding) next to his Jacuzzi. And Marin County had such a skunk problem that the local humane society offered a seminar entitled "Skunks: Whose House Is It?"

Skunks pose four problems for homeowners. First, they spray. While most people know enough to keep away from skunks, they're not always successful. And cats and dogs don't necessarily even know to stay away. Second, skunks can spread rabies through their bites. Third, they dig up lawns, looking for insect larvae. Fourth, they enjoy rummaging through garbage left outside (or partaking of vegetable gardens).

A lot of people are frightened at the prospect of being sprayed, and with good reason. The smell from a skunk's spray can reach up to a mile, and up close, say in a house, it can make folks feel nauseated. Besides the smell, the liquid causes eye inflammation. Fortunately, there are some precautions you can take to avoid being sprayed.

The first thing is to be alert if you discover—say, during a walk in the woods—that you're in skunk territory. Like backpackers who've been in the woods for a while, skunks are often smelled before they are seen. But if you don't smell them, you might hear them. While they're basically silent creatures, you might notice them grunting when they feed. When agitated, the skunk may scold, growl, or snarl. If you come across a skunk, don't advance. There's a chance it will try to get away. Be even more wary with baby skunks, which are more likely to spray than adults. So don't walk up to a tiny skunk and say, "Oh, how cute, a baby."

What can you do if a skunk sprays you? First, don't plan a big date for that weekend. A bath in tomato juice is one prescribed course—a fairly expensive remedy, but it will reduce the odor. After your tomato juice bath, you might consider taking a shower so you don't smell like tomatoes. Rubbing yourself with diluted vinegar is another prescription. The most effective treatment is a liquid form of neuthroleum-alpha, available from hospital supply shops. Wash with the solution to get rid of the skunk smell. Also, spray to the eyes may cause temporary blindness; rinse them with plenty of water. Sight should return in ten or fifteen minutes.

As for your clothes, they may well be a lost cause. But if you want to try to salvage them, you can wash them in laundry detergent and household ammonia or bleach. It helps to let them air out as much as possible. Trapper Janice Henke of Glenn Falls, New York, says that when a skunk sprays her, she takes off "every stitch," including her sneakers, and hangs everything in a tree. "I leave my clothes outside for a week or so." The utility of this remedy might depend on how close your nearest neighbors are and how much you like them.

Dogs and skunks are another issue. Dogs like to chase smaller animals, and there's nothing you can do to teach a dog that skunks are meant to be avoided—at least not until it's too late. Then, when the inevitable happens, a smelly dog may only be the start of your problems, as the Minneapolis owners of Ava discovered.

The Backyard Feeders

One morning, Ava went out just before sunrise for a little run around the yard and found a skunk to chase. As you know, a skunk's aim is nearly perfect every time. While Ava's problems couldn't have gotten worse after she was hit, her owners' problems had just begun. Ava decided that to get rid of the smell on her fur, she'd rub it off. So, naturally, she ran into the house and rubbed ferociously on every piece of furniture she could find. If your dog is skunked, use the same treatment as for humans: Give it a bath with tomato juice, diluted vinegar, or neuthroleum-alpha. Vermont resident Pam Dennis used peppermint mouthwash to cleanse her dog of skunk spray. You might also try a fresh-scented perfume to mask the odor until it fades.

People (and to the degree possible, their pets) should also avoid being *bitten* by a skunk. Greater numbers of skunks are becoming infected with rabies, which can be fatal to humans if not treated. In some areas (most notably the eastern United States) the number of rabid skunks now exceeds the number of rabid raccoons. The increase in rabies among skunks has been attributed by some experts to the decrease in interest in skunk furs among clothing buyers. The fewer furs bought, the fewer skunks trapped, and the greater the increase in the skunk population and the greater opportunity for rabies to spread. Yet, many biologists don't believe that trapping dampens the incidence of rabies among the population.

Rabid skunks may act aggressively, or they might not. They have been known to attack dogs and cats during the daytime. Perhaps one sure way to identify a rabid skunk is that a skunk with rabies won't spray (if that's any consolation). All they want to do is bite.

A number of cities have started programs to vaccinate skunks against rabies. The vaccine is 100 percent effective on laboratory animals, but less than 80 percent effective against wild skunks because of the effect of weather and temperature on their metabolism. But the vaccine's effectiveness goes beyond these biological statistics. Because skunks are territorial, one family of inoculated skunks will fend off other, non-vaccinated ones.

A skunk's effect on your lawn is possibly irritating, but less worrisome. Because skunks seek out grubs (insect larvae) that are relatively close to the surface, these critters don't make the kind of far-reaching holes that gophers do. They only leave a series of small, two-inch-deep holes in your grass. Of course that's no consolation to anyone with a newly sodded lawn. But if you can bear it, nature automatically repairs skunk holes during the wet spring months. Indeed, some homeowners find that skunk holes are more desirable than the insects they eat.

But some homeowners don't want either grub worms or skunks. In that case, one strategy involves eliminating the skunk's food source. *Bacillus popillae*, milky spore disease, will get rid of the grubs and the feeding skunks. (This solution was mentioned earlier in the section on insects. The parasitic nematode strain Hh, also described earlier, will kill the grubs more quickly.) Another way to keep the skunks (and raccoons) out of your yard is not to fertilize your lawn. Fertilizers make grass lush, which attracts grubs. Cultivate a slightly less-lush lawn and the grubs, skunks, and raccoons will probably dine elsewhere.

At the same time that you're attacking your local grub population, you need to thwart the skunk's access to other food sources. Don't feed your pets outdoors, and secure your garbage. Put food in plastic garbage bags, place the bags inside cans, and then cover the garbage cans securely.

To prevent skunk problems, you also need to remove potential shelter areas. Pile firewood on pallets that don't rest on the ground. Remove brush and piles of leaves that skunks might use for denning. They may be coming around to feed on mice in the yard, so get any mouse problem under control. Put wire mesh in front of drain openings, and use the same material, wire mesh, metal sheeting, or concrete to block crawl spaces under porches and buildings and any other holes that skunks might use to enter a house. They won't chew through wood. Skunks also don't climb, so don't worry about them getting to the upper reaches of the house.

For dens, they like junky areas, outbuildings, garages, or holes under your house or in your yard. What you will need to do is block off the entrances to the den—without trapping the animal in the structure. (It's inhumane to the animal and also unsafe for you.) Be careful between May and August when the mother skunk is likely to have her young in her den.

When the den is under your house, first seal off all possible entrances to the den but one. If you're sure there are no animals in residence, then block that hole as well. To find out when the animal has left the den, sprinkle flour on the ground in front of the entrance, extending it about two feet in all directions. Examine it periodically after dark for evidence that the skunk has left to feed. After the den is empty, close the hole. If you don't find tracks, repeat the process for a few days.

If you don't want to do midnight repairs, you could build a one-way door that swings only outward, but makes sure the skunk doesn't get back in. Skunks aren't dexterous like raccoons, so they'll have a hard time lifting the door. Again, do not use this stratagem in the late spring or early summer, especially during May to August, because baby skunks may be trapped inside.

If the skunks live in a hole, culvert, or similar area, an alternate approach is to try flooding their space with water. It will probably drive them out, although the skunks will announce their displeasure while leaving. Also, if you resort to flooding, be careful there are no babies that would be drowned. Mothballs or ammonia-soaked cloth will also drive them out of a den or keep them from visiting an enclosed area.

If a skunk wanders into your home or other building, it's best to let it leave on its own. Open all the windows and doors and leave yourself. Peggy Pachal, a canine behavior consultant with the American Dog Owners Association, once had a skunk enter her home while she was working outdoors. Her infant lay sleeping on the floor of the house. "I was afraid the baby would start crying and startle the skunk." Thinking quickly, she opened a can of dog food and put it on the porch to lure the skunk outside. It worked, and she entered the house from another door and closed the door behind the skunk.

Occasionally skunks get trapped in window wells or other pits. You can carefully lower a board into the pit so they can climb out, and you can make the escape easy by taking the time to attach footholds, nails, or cleats to the board at six-inch intervals. Approach the skunk slowly, talking softly—don't do anything suddenly. Often skunks won't spray when treated gently.

Alan and Mary Devoe relate a charming skunk tale in their book *Our Animal Neighbors*. On a walk one spring evening, they came across a skunk that had his head trapped in a tin can. Talking softly to the skunk, Alan knelt down beside it and actually petted its back. The can was tightly fitted to the skunk's head and Devoe had to grasp the skunk and firmly pull the can away, working it back and forth, and twisting it around. Finally the can came free and Devoe slowly stood up. "The skunk sniffed at our feet, brushed his body and tail back and forth across our legs as a happy cat will do. . . . He sniffed a time or two at the snow-filled air, turned from us, and went rocking off in his crooked tracked way," they wrote.

If you have a skunk problem, you might also ask your local wildlife officials to trap and remove the animal. As with many wild animals, it's not a good idea to try to trap a skunk yourself. Many are rabid, and skunks often manage to bite the person who trapped it during transport and handling.

If you decide to capture a skunk on your own, carefully follow the directions for caging it. Think ahead to the release and attach a string or wire so you can open the door from a safe distance. Also cover the trap with canvas, leaving only the door exposed. Then when you get the animal trapped, the skunk will feel more secure. Failing that, when the skunk is trapped, approach the cage slowly and quietly and then gently cover the trap with a canvas so the animal stays calm. It will feel more secure in a dark environment. Carry the trap gently. Transport it ten or more miles away to a suitable habitat. (I think a truck is the best method for transportation; the canvas is no guarantee that the skunk won't spray.) Good bait for a live trap includes fish-flavored cat food or bread with peanut butter or sardines.

There is one possible way to keep a trapped skunk from spraying. This method is not 100 percent effective, but if you're a gutsy type and not planning

---

CONTROLLING SKUNKS

- Block access to crawl spaces under the home to prevent them from entering.
- Clean up garbage and eliminate access to pet food.
- If you encounter a skunk, try to act calmly and move slowly to get away from it.
- If you get sprayed, try tomato juice, neuthroleum-alpha, diluted vinegar, and plain soap and water for your body. Use detergent, ammonia, bleach, and fresh air for your clothes. And apply tomato juice, neuthroleum-alpha, diluted vinegar, mint mouthwash, aftershave, and/or soap and water for your pets.

---

on going out for the evening, it might be worth a try. Once you have the skunk trapped, slowly approach the cage, holding a sheet of plastic in front of you. This way, the skunk may not know that a person is approaching. The plastic could also protect you from a direct hit.

Spread the plastic over the cage and turn the cage so that it's vertical and the skunk slides to the bottom. Then use a plunger-shaped object to push the skunk into a corner. Once squeezed in place, the skunk won't spray. Although this technique sounds both complicated and risky, it's used by some professionals to capture and vaccinate skunks.

Finally, a word on a way *not* to get rid of a skunk: Shoot it at close range. This is something that people may only try once. As soon as it's shot, a skunk will give off a spray.

PORCUPINES

One of the largest rodents in North America (after the beaver) is the tree porcupine (*Erethizon dorsatum*), which can weigh more than twenty pounds. They average about eighteen to twenty-four inches, which includes a tail that may be about six inches long. But the feature for which they are best known is their quill-covered body. With the exception of the face, underbelly, and tail, their bodies are covered with rigid spines. These spines are really just hairs that are loosely attached to their skin and an adult porcupine has about thirty thousand of them on its body. And, contrary to the myth, porcupines can't "throw" their little darts.

Their quills enable them to challenge their enemies, because most predators are not brave enough to tangle with these chunky little fortresses on four feet. Nevertheless, some of the fiercer and more cunning forest animals know how to attack them relatively safely. Because the quills are loosely attached to the skin of the porcupine, they easily dislodge when the other animal makes contact; however, the "business" end of the quill has a nasty little barb on it. Once these are imbedded in the skin, they are extremely difficult and painful to remove. They may also cause infection. This can eventually kill the predator, or it may starve if it can't eat properly due to injuries. Anyone who has had the unhappy job of de-quilling a dog's face knows just how unpleasant a porcupine encounter can be. Meanwhile, any missing quills on the porcupine will grow back within a few months.

One of the porcupine's few feared enemies is the fisher. A fisher (*Martes pennanti*) is one of the larger members of the weasel family; it lives in trees and forages at night—just like the porcupine. And it is the one mammal that really knows how to deal with the "quill factor." It attacks the porcupine right in the face, making it either bleed to death or sometimes simply killing it from shock. Or it manages to flip the porcupine over on its back so the fisher can attack its unprotected belly. The fisher also has swiveling ankle joints, so it can climb up and down trees quickly—certainly faster than the porcupine. When the fisher population goes down in a forested area, the porcupine numbers—not surprisingly—go up.

As suggested by their name, tree porcupines enjoy eating trees. But not just any trees. They do have a few favorites, including spruce, elm, and hemlock, and they will systematically kill a tree by removing the outer bark layer. This exposes it to insects, disease, and harsh weather. In order to do all of this chewing, nature has equipped the porcupine with extremely sharp and strong teeth, as well as a tough neck and skull. And they have short, strong legs for climbing.

During the day the porcupine tends to rest, and is often found lazing high up in the fork of a tree, curled up in a ball. In the summer, it feeds mostly on the ground or on the lower parts of trees. In the winter, they climb up deciduous ones (maple, beech, or oak, for example) to feed on the buds and the bark. They may also eat the needles of coniferous trees. Active all year round, a porcupine is a hardy animal that manages to live about ten years in the wild.

If you have anything made of wood in your yard, and porcupines are in the area, you are at risk of having your tools, furniture, and even your back deck destroyed. Porcupines love salt, and human perspiration, which is rich in salts, gets on the wood and makes it especially attractive to them. Canoe paddles—repositories of sweat—are also favorite targets. In fact, porcupines have even been

CONTROLLING PORCUPINES

- Add wood preservatives to your deck and outdoor furniture to make it less tasty and appealing.
- If trapping is required, use foothold or box traps. And be sure to check them regularly.
- Use wire mesh or aluminum sheeting to "wrap" vulnerable young trees and protect their bark.
- Keep paddles, axes, and other wooden items properly stored. Leaving them outdoors at a cottage simply invites porcupines.
- Put towels on your wooden decks when you are sunbathing. That will keep the sweat (and salt) off the wood.
- Keep your dog on a leash if there are porcupines in the area. Young dogs—who haven't yet learned the quill lesson—are especially vulnerable.

known to gnaw on car tires in areas where the roads are salted in the winter. They also seem to like plywood, possibly because of the glue used between the laminated layers. Porcupines will feed on rosebushes, pansies, and various crops in vegetable and fruit gardens, too.

These slightly odd and unattractive creatures tend to be solitary most of the year, except when it comes to mating season. Copulation is done delicately and quickly, in a "hands-free" maneuver. Unlike most of the rodents examined in this book, female porcupines give birth to one litter in the spring annually, which usually produces only a single offspring. Their gestation period is also uncharacteristically long for a rodent: seven months. With all of those quills, birthing could be a dangerous event. Once again, nature offers clever protection. The young are born headfirst, and they come out in a placental sack. At this stage, their quills are soft, but within a few hours they are stiff and able to stand erect.

The greatest hazard to a porcupine, and the cause of most deaths, is car traffic. During the summer months, they tend to join raccoons as the two most frequently killed animals on our rural highways. They don't move quickly, and they aren't easy to see at night. Fortunately, they don't leave a smell, as do skunks when you hit them, and their quills won't puncture your tires.

The Backyard Feeders

SQUIRRELS

In the early 1900s when hunting and deforestation were taking their toll on wildlife, there was some concern that gray squirrels might become extinct. They've done a pretty good job of making a comeback, though. If we look at sheer numbers, it's hard to find a more widespread critter in North America.

If you're wondering whether the gray squirrel is actually gray, the answer is no. Its fur is a combination of gray, brown, and yellow on top, and white and gray on its underbelly. Its long, soft, bushy tail, which is about eight to ten inches long, is silver and white. This tail is actually a source of warmth and provides balance, which it needs when it's scaling trees and landing on your bird feeder. Squirrels also use their tails to communicate with one another. The gray squirrel, *Sciurus carolensis*, is about sixteen to twenty inches long and can weigh up to twenty-five ounces. Gray squirrels produce two litters a year, but they only live for one year on average. Domesticated squirrels can live up to twelve years, which can mean only one thing: Life on the streets is tough.

There are many species of squirrels, but they are usually divided into two groups: tree-dwelling squirrels and ground squirrels. Gray squirrels are tree dwellers, which means that they pose a serious threat to your nut trees and bird feeders. Usually they prefer beech and oak trees because they can dine on their nuts. In addition to exploiting nut trees and bird feeders, they also indulge in fruit, flowers, insects, bird eggs, peanut butter, apples, plant buds, and mushrooms. Baby birds are also part of their diets, which is no surprise, since gray squirrels are omnivorous.

Gray squirrels can be found in several areas of North America, but if you reside in the eastern region, then you've definitely been exposed to them. Areas heavy in vegetation, such as parks and forests, are some of their favorite habitats. Gray squirrels are forward-thinking creatures, so they bury nuts during fall with the intention of going back to them later in the winter. The reality is that most squirrels can't "remember" where they buried their little treats. They simply find them again by using their keen sense of smell.

Gray squirrels are also known for stripping bark off hardwood trees. As a result of this pastime, they are commonly referred to as "tree rats." Gnawing and stripping a tree of its bark can kill the tree or make it susceptible to disease. In this way, they are similar in habits to the porcupine, and they're no friends of tree planters.

Tree squirrels, named before anyone knew about their passion for homes and bird feeders, are different from ground squirrels, which are cute animals, but with none of the cunning and agility of their arboreal kin. Either variety, however, may attack your garden. Tree squirrels, true rodents, are also known region-

Gray squirrel (*Sciurus carolensis*)

ally as fox, eastern gray, western gray, tassel-eared, or pine squirrels. They are all daylight squirrels, although their more exotic cousins—the flying squirrels—are nocturnal creatures.

Tree squirrels range from coast to coast across North America. The largest are the fox squirrels, which are eighteen to twenty-seven inches long and weigh up to two pounds. The smallest are the pine and flying squirrels, which are eight to fifteen inches long and weigh up to twelve ounces. About half of a squirrel's length is actually its bushy, flattened tail, which is usually held in an S-shaped curve over the body while in a sitting posture. The hind legs are larger and stronger than the front ones and are used when leaping from tree to tree. The front feet are adapted to holding nuts while sitting in an upright position. Squirrels come in colors ranging from black to gray to red.

Squirrels are quick and nimble. They can run, climb, and jump among the branches and twigs of the loftiest trees, executing risky arboreal acrobatics. People who tend to hyperbole say that before this country was settled, a single squirrel could travel treetop to treetop from the Atlantic to the Mississippi without once touching the ground. When startled on the ground, a tree squirrel usually scrambles up the nearest tree, traveling swiftly from one to the next, seldom missing a foothold. When it exceeds its ability to jump, it may drop from the tree to the ground and scurry up another tree, apparently none the worse for any injury. Squirrels navigate in lanes in trees, called "travel lanes," which they mark by scent. They will also travel from pole to pole along electrical and telephone wires, and occasionally on trees and houses, for city blocks without setting foot on the ground.

Squirrels have home ranges, and their sizes depend on the availability of food. Typically, a home range varies from one to seven acres. If the population is especially high, gray squirrels have been known to migrate in search of new habitat. And they are extremely territorial. You'll hear about it if one of them is annoyed by your presence on its turf. It will express its dissatisfaction with an extended vocabulary of slow warning barks, scolding, teasing, and assorted playful *chucks*. When alarmed, or when another squirrel is approaching a food store, jerky tail-flicking usually accompanies these calls.

Before bird feeders arrived on the scene, squirrels subsisted on what they found in their natural habitats. They depend on what are called "mast" crops: acorns, hickory nuts, and beechnuts. In many parts of North America, squirrels busy themselves with burying nuts during the late summer and early fall. Although caches of nuts are sometimes stored in buildings or hollow trees, the nuts are usually stored singly. Throughout the winter they dig up these cached or buried nuts for food. When the mast crop (oak and beech nuts) is poor, squirrels are hard-pressed to obtain sufficient food. Food shortages, combined with severe winter conditions, may reduce populations and result in poor reproduction.

Mature squirrels can consume up to two pounds of nuts per week. They may travel extensively in the fall to locate food if supplies in the home territory are low. Since nearly all forest wildlife species use mast, there is a direct competition for food between squirrels and other animals such as ruffled grouse, deer, black bears, chipmunks, white-footed mice, blue jays, flying squirrels, turkeys, and others. This competition points out the complex relationships of the forest wildlife community.

When spring thaws occur, gray squirrels utilize new foods as they become available. Swelling buds and flowers of red and sugar maple are eaten in April. A little later, the winged fruits of red maple and American elm may be consumed. Summer foods consist of berries, mushrooms, apples, corn, and other grains. Squirrels may occasionally eat birds' eggs and chew on bone and deer antlers for calcium, phosphorus, and other necessary trace minerals. Flying squirrels share tastes with their cousins, but they eat more meat—bird eggs, birdlings, and insects. Pine squirrels eat the same menu, but as their name implies, prefer coniferous forests where pines provide most of their food. All squirrels will cache nuts for later consumption.

Squirrels are active year-round, although they may stay in the home den for several days during storms or severe winter weather. In winter, several unrelated squirrels may share the same nest. During the rest of the year, the periods of greatest activity are at dawn and in late afternoon. Although wind discourages their movement, squirrels will feed in rain or snow if the wind is not strong. Squirrels are somewhat gregarious, tolerating each other in small groups except when food supplies are low.

Squirrels use two kinds of nests: leaf nests and tree dens. Good tree dens are permanent quarters, while leaf nests serve as temporary homes during the summer. They prefer cavities in mature living trees for winter dens. Den openings are frequently formed where branches have fallen off or feeding woodpeckers have drilled holes into trees. Weather, rot, and insects begin to hollow out cavities at these locations. Gray squirrels must periodically gnaw back new callous bark tissues to keep den entrances from sealing over. Den openings are usually about three inches in diameter. Old hollow trees with broken tips, cracks, and many openings do not make good tree dens, but do provide hiding places for gray squirrels. On occasion, squirrels may den in barns, garages, or attics.

During the summer months, adult squirrels frequently build and occupy leaf nests, which are usually built in the top fork of a tree or in the crotch of a limb near the trunk. Most nests are generally globular in shape, but their form may vary depending on the supporting base. A single entrance usually faces the main tree trunk or the nearest limb that provides access to the nest. Three or four structural parts make up the nest: the base and support are made of twigs from the nest tree, the floor of the inner chamber is made of a layer of compact humus and organic debris mixed with twigs, and the outer shell of woven bark, grass, and leaves may be constructed to provide warmth and added protection. The outside diameters of these nests range from fourteen to sixteen inches, and they weigh from six to seven pounds. Leaf nests may serve as temporary quarters near seasonal food supplies. They are also cooler and free of the parasites that accumulate in winter dens.

Squirrels mate once a year, except the fox and gray squirrels, which mate in December, January, or February, and again in the early summer. One reason squirrels are so difficult to breed in captivity is that a female's fertility seems to be boosted after being chased helter-skelter through the trees by one, two, or more males. After a forty- to fifty-day gestation period, the female gives birth to three bald, blind, earless offspring. The young are dependent on the female parent for nearly six weeks; their eyes do not open until around the thirty-seventh day. Young venture from the nest or den at eight to ten weeks of age. They climb about the den tree for a few days before actually descending to the ground for the first time.

Spring litters are usually born in a tree den, while summer litters may be born in leaf nests. Many times, the number of females producing a second litter depends on the availability of food. Squirrels (especially males) do not keep mates for life, and may migrate between litters. As a consequence, genetic inbreeding among squirrels is rare.

Natural predators of the squirrel include humans, hawks, owls, foxes, bobcats, and raccoons. House cats also prey on young squirrels. In the fall, when squirrels are prone to migrate longer distances in search of food, the roadkills are usually high. The average life span of gray squirrels is eighteen months, but they have been known to live to eight years. Mortality is high in the wild, and about half of the squirrel population dies each year. While squirrels can live up to twelve years in captivity, a four-year-old wild squirrel is elderly. Although many animals like to eat squirrels, predation doesn't have much of an effect on the population.

They are also fair game in many parts of North America. Plenty of people hunt squirrels, even folks without a garden grudge who just like squirrel meat. The drawback? "They look like skinned rats," says Barbara Meadows, wife of a dedicated West Virginia hunter who lovingly fills her sink with the rodents. He has to cook them himself, though.

Squirrels are notoriously tough to control. Bill Adler Jr., author of *Outwitting Squirrels*, confirmed the need for more than haphazard techniques while he was being interviewed for a television show's special segment on squirrels. So that the interviewer could photograph squirrels in action, he and Adler started walking around the yard, strewing nuts to attract them. As they strolled, Adler noticed one squirrel scurry up to the open back door. It had sauntered right into his kitchen!

He raced up the deck stairs screaming for his wife (who was at that moment tending to their screaming baby, so she noticed nothing). The cameraman raced up behind him, not intending to help, but to film the episode for television. Luckily, Adler managed to shoo the squirrel back out the open door—a trick that was more a matter of luck than skill. He knew there had to be better ways.

There are two schools of thought about squirrels taking up housekeeping in your home. One is to be ruthless and assume that a squirrel will always remember what a lovely home you have and attempt to return. The other is to be patient, believing the squirrel will eventually give up and live in a tree.

TOTAL CRITTER CONTROL

John Adcock, a pest control professional, says, "You can't chase squirrels out of your house. Squirrels have to be trapped. They'll never leave!" He bases his advice on experience. Adcock removed a fascia board from a house in Chevy Chase, Maryland. It had thirty-seven holes in it, and that was just one board. There were four fascia boards on this house which represented fifteen years during which the homeowners had dealt with squirrels by simply chasing them out and closing up the holes. "It doesn't work; they just chew another hole in the house. They have a vested interest in the house, due to the fact that they have stored food there. This is where they're planning on raising their young," Adcock sums up. All without having to get a mortgage!

And there are better and worse times for plugging up squirrel holes. Sometimes folks patch up a squirrel access hole only to find a crazy mother squirrel that's chewing holes elsewhere in the house. They throw stones and yell, and the squirrel keeps it up. Yet imagine if someone blocked off the door to *your* nursery—you'd go through anything to get to the baby. A squirrel mother is the same way. Get that access opened up, let her get to the babies, and she'll be all right.

Don't fix things after the fact. The time to screen is *not* during the middle of a squirrel infestation. And don't declare victory too soon. In the summer, attics can reach temperatures of 120 or 130 degrees Fahrenheit, and no squirrel in its right mind is going to go up there. When it gets hot like that, people think the squirrels have moved on and there's no need for repairs. In reality, the squirrel was ready to leave that area anyway. Come next September or November, however, they're back there hammering on the house, chewing another hole. But now the guy who had one hole in his house has two, because of the fact that he never took care of the original problem. The first priority is to get rid of the squirrels that have the interest in your property. They consider it their home.

Becca Schad of Wildlife Matters concurs: "You get whole generations of squirrels who were born in attics and think that's where they're supposed to live." Compared to a warm, dry attic, a tree nest looks flimsy; the wind could blow, it could fall out. But squirrels are probably the most destructive animals that nest in attics. They gnaw everything: insulation off wires and boxes, stored clothes, and your tax records, too.

Schad represents the patient school of squirrel control. Like Adcock, she would never attempt to oust a mother squirrel, but once the babies are grown, she becomes ruthless, too. She blocks up every hole. And she cuts back branches that allow squirrels to get to the roof or attic. John Hadidian of the National Park Service says, "Squirrels may be scared away with strips of Mylar tape on the fence or a rag soaked in ammonia."

Outdoors, squirrels are just as tenacious. "What happens often is you get populations that have a tradition," says Hadidian. "You get tomato-eating squirrels, and the mother teaches the young that this is a food source. Three blocks away, the squirrels don't eat tomatoes. You have to teach the squirrels in your neighborhood that tomatoes aren't food. But you ought to be able to do that and you ought to count on taking two years to do it."

Hadidian points out that mothballs, often used indiscriminately outdoors, are systemic compounds that are absorbed into plants. Dried-blood fertilizer may do a better job outdoors, and it will nourish the plants, too. Used indoors in enclosed spaces, mothballs or flakes may encourage animals to leave long enough for you to close off a point of entry.

John Adcock has a different view: "Mothballs are camphor. Camphor is actually flammable. To put camphor in an area that gains temperatures of 120 to 140 degrees Fahrenheit is ludicrous, as far as spontaneous combustion is concerned." He wonders how many mothballs are necessary in a two-thousand-square-foot house with nine hundred square feet under the roof. "What do you do? Put a mothball every six inches? If you do, they just kick them out of the way. You'd have to fill that whole attic full of camphor to make it work."

Rather than use camphor, some homeowners turn to ultrasonic devices to rid their houses of squirrels. While many experts say that ultrasonic devices don't work, manufacturers and retailers insist that ultrasonic devices offer effective rodent control. Lowell Robertson of Sonic Technology, a California manufacturer of Pest Chaser ultrasonic devices, illustrates their efficacy with a tale from his own attic. "The devices seem to work really well on squirrels," he says. "I'll tell you straight off that we have not done specific testing on squirrels, but I know from my own experience that it got squirrels out of my attic. Squirrels can apparently perceive a portion of the ultrasonic band that we're broadcasting." He found that tenacious squirrels were gnawing the wood in his attic:

> Squirrels can eat right through a two-by-four. When you exclude them from a nesting area they like, they just chew another hole. I literally pounded up twenty square feet of hardware cloth at all the points where I had evidence of them getting in. And they just chewed another hole. That's when I finally put one of my own products in my attic. And that cured the problem. Many people have said that they have had the same experience with the product.

So, do sonic devices work or not? Well, they seem to be like elephant whistles. You've heard the story about the New Yorker who blows on an odd-sounding whistle. His friend asks, "What's that for?" The whistle-blower replies, "It's to keep elephants away." His friend, incredulous, scoffs, "No way can that work." To which the whistler says, "But do you see any elephants around?" Some people insist that sonic devices *must* work simply because no one has definitively proved that they *don't*.

Discovering which control method works for you may entail a lot of trial and error. In the meantime, though, squirrels can be annoying—but at least they're not exactly life threatening. Diane White, a writer for the *Boston Globe*, and her cat were helpless against a female squirrel. It entered the home through the cat's door and helped herself to English muffins, croissants, and chocolates. While White's husband suggested a BB gun solution, she advocated the do-nothing approach, not wanting to keep her cat from the litter box on the other side of the swinging pet door.

Whether they're in your attic or in your yard, the only "final" solution may be to trap and relocate an especially pesky squirrel. Garon Fyffe, director of ABC Humane Wildlife Rescue and Relocation near Chicago, relocates about eight hundred squirrels a year, moving the varmints from where they're unwelcome, and taking them to spots where they're wanted. He uses live-traps and often moves them to wildlife preserves, where he has a permit to release animals. He's careful to make sure the preserves don't abut neighborhoods where the squirrels will become problematic to someone else.

It's easy enough to capture the little critters and take them for a little country ride to an oak tree stand, but how do you know they won't just return? The answer is that squirrels migrate to some extent, but they're not homing pigeons. They probably won't return to your home. We've all heard stories of cats that've found their way back from a hundred miles away after a mishap, but squirrels apparently can't duplicate those feats.

Squirrels *do* have their limitations, but when you are trying to cope with the ones in your yard, it may seem as if no feat is impossible for them. A Canadian naturalist and animal rights activist, Barry Kent MacKay, decided he had met a new squirrel species, *Sciurus carolensis hellenis*, the gray squirrels from hell. They attacked his flowers, his birdseed, and his bird feeders, "chewing through plastic as if it were compressed peanut butter." He tried everything: squirrel baffles, squirrel-proof feeders, greasing the bird feeder's pole, moving the bird feeder pole, installing squirrel deterrents on the feeding platform, and cutting back branches on trees all around the feeder—until "the yard looked like a war zone." Eventually he managed to keep the squirrels on the ground, where they subsisted on the dropped seeds from the birds, but he now knows it's only a matter of time until the demonic animals figure out a new access to the seed supply.

In other parts of North America, there are less tenacious squirrels. Often, greasing the bird feeder's pole works to keep them grounded, especially if you mix Vaseline and cayenne pepper or curry powder. Other greasy concoctions—including axle grease or solid vegetable shortening—are *bad* ideas. Axle grease is lethal, and the shortening often spoils and will sicken squirrels and other animals. (It might also attract ants, too.)

The Backyard Feeders

Generally, squirrel-proof bird feeders are selective. There are feeders that only open when a specific bird alights on the perch. You set the dial to the bird you want to feed, and a system of counterweights opens only to a bird of that weight. Since birds of a species weigh more or less the same, there will be no anorexic starlings eating from the feeder you've set for a goldfinch. Other feeders use wire to keep squirrels away from the feeding tube. Pick one that won't allow a squirrel to get its teeth against the plastic or it will simply eat the feeder. Another model that's virtually squirrel-proof is the MH Industries GSP (Guaranteed Squirrel Proof) Feeder. Only animals that can fly (and small helicopters) can get inside the globe that opens at the bottom. The feeders that operate by closing a door in front whenever something heavy lands are effective, too. These go by names such as Absolutely Squirrel-Proof Feeder, Squirrel's Defeat, the Foiler, and the Steel Squirrel-Proof Feeder.

Try hanging the feeder on a small chain the squirrel can't chew, and install a baffle above the feeder. Some squirrels aren't especially brainy. Robert Dewey of Washington, D.C., baffled squirrels with no more than a metal coffee cup strung onto the hanging wire. (No, it doesn't have sharp edges.) But it was enough to convince the squirrels to feed on the ground spills with the mourning doves.

When all else fails and you can't beat 'em, you might just feed 'em. That's what some people eventually decide to do. Give the squirrels a feeding station of their own, and maybe they'll leave the birds alone. Or at least they'll show a preference for food that's easier to get at, and they won't resort to chewing up your custom bird feeder.

Despite all of the various predators in and around a city, there are only a few that hunt squirrels. In fact, the car is probably the main cause of death for these critters. It's not that squirrels aren't fast and wily, they just have bad depth perception and close-up vision. Add to that the confusion and noise of traffic, and you have a plausible explanation for the squirrel carnage on our roadways. The average number of squirrel roadkills is one per every ten miles of pavement.

The more rural and slightly more "attractive" variety of this rodent species is the red squirrel. These trim little critters, *Tamiasciurus hudsonicus*, are more petite than gray squirrels, weighing in at nine ounces on the high end. Their coats are usually a brownish red with a white or off-white underbelly, while a white ring encircles their black eyes. The diet of the red squirrel does not differ from that of gray squirrels, except that in addition to seeds, fruits, berries, mushrooms, and nuts, red squirrels like fungi.

Red squirrels are especially fond of pinecones. They are very quick creatures and often can be found darting around on the ground or along coniferous branches. During the fall they zip about collecting seeds and nuts for winter, and during the winter they run about furiously looking for buried food. The red squirrel is more compulsive about having food in reserve than the gray squirrel.

Even if they have plenty of food buried in a given area, they are constantly searching for more.

These squirrels are found across North America, but you're more likely to encounter them at a cottage, cabin, or chalet in the country. They tend to be the "chatty" ones, scolding you from high up in the white pines. Or they may be bolder and try to steal food off your cottage deck. Because they are "cuter" than other rodents, we are often tempted to feed them. As suggested before, that's a bad idea; once they've been encouraged, it's hard to keep them at bay. Red squirrels produce litters twice a year, with each one averaging five offspring. If you don't want hordes of these little chatterers and nut-storers invading your space, don't encourage them.

Finally, if you've arrived at the place where the war is either over or you've admitted defeat, here are some amusing squirrel distractions. A "three-ring circus" or "squirrel-a-twirl" is a three-armed windmill with an ear of corn attached to the end of each arm. Mount the contraption on a tree or any flat surface. The device rotates as the squirrels try to get the corn—and they will certainly try. If that's not silly enough, you can get a table and chair set just for Squirrel Nutkin. An ear of corn stands upright on a spike on the table, and the squirrel sits in the chair as it feeds. Double your fun with a place setting for two!

CHIPMUNKS

If you take a quick look at a chipmunk (family Tamias), you might mistake one for a baby squirrel, because they look very similar. They are usually tan with white-and-dark stripes down their backs. And like squirrels, chipmunks love to nibble on seeds and nuts. They range in size from three to eight inches in length and weigh up to five ounces. You probably have noticed that many of the names of rodents are directly linked to one of their identifying characteristics; for example, the muskrat was named for the strong odor it emits. The same rule can be applied to the name of the chipmunk. Chipmunks are very vocal at times, and it is nearly impossible to ignore the unique noise they make, which sounds like a *chip*. Chipmunks usually make this noise to notify others of impending danger. They also seem to have a chippy countenance and are known for the abrupt manner in which they dart around.

Unfortunately, these chipper critters don't dart past your nut and fruit trees without cleaning them out first. If it were up to rodents like chipmunks and squirrels, you wouldn't have one walnut to harvest when the season was over. This is because they prepare for the barren winter months by pillaging your yard. But the most damage occurs when chipmunks make a run for your flowerbeds.

Garden bulbs are at risk of being eaten, so keep a close eye on your flowers if you know that chipmunks are in the area.

OPOSSUMS

The opossum that is commonly found in North America resembles an oversized rat. Its head and body measure about fourteen to twenty inches, and its long, rounded, and almost hairless tail may add another ten to twenty inches. It has a long, pointy, pinkish nose, and a face that is mostly white and gray. But being marsupials, a female opossum will have a feature that a rat doesn't have: a pouch. This "pocket" allows a mother to carry her young, where they nurse for a three-month period. Her litter can range anywhere from five to twenty offspring, and they stay in her pouch for a few more months while they nurse. Then they get to ride around on her back for a few more weeks.

Opossums are, in fact, the only marsupials in North America. That makes them not-too-distant relatives of the kangaroo, another member of the "pouch" family. Like porcupines, opossums are thought to have originally lived only in South America, but when the two continents were joined (probably by volcanoes), they migrated north. In South America, there are more than forty-five varieties of these unusual creatures. In North America, there's only one, *Didelphis marsupialis*, the Virginia opossum. They've now grown accustomed to North America, but they have an advantage in adapting because they're omnivorous. And apparently, they're reasonably bright, even though their brains are only one-sixth the size of a raccoon's.

There are a few opossum features that haven't quite worked in the transition from one continent to another. They have naked ears and naked tails, which means that they have real problems with frostbite in the cold weather. Some opossums actually get gangrene from these encounters with the snow and cold. They also don't hibernate or store food externally, so they've clearly fared better in the southern United States than in the more northern regions. Until Europeans arrived in North America, apparently the opossum didn't exist north of Virginia and Ohio. Once their predators (cougars, fishers, wolves, foxes, and coyotes) were hunted extensively, opossums began to thrive and even extend their territory. They were even introduced to southern British Columbia as a small game mammal around 1950.

Opossums like to be near water, and prefer farms and forests. In that way, they pose a threat to chicken coops as well as gardens. Their diet also includes grain crops, fruits and vegetables, small animals, birds' eggs, a variety of insects, mushrooms, and even dead creatures in the woods. Opossums have fifty teeth—a number greater than any other mammal. They may look slow and slothful, but they are well armed.

Opossums eat moles and shrews—animals many other predators don't like because of the musky smell of their meat. They're known to eat snakes, too, including vicious rattlers, water moccasins, and copperheads. (Apparently, opossums are immune to their bites and venom.) With opposable toes on their hind legs, they're particularly good climbers—which helps them escape enemies and find food in trees. In order to get through the tougher and leaner winter months, they can also accumulate up to 30 percent body fat.

They tend to be shy night creatures, so opossums aren't often seen in the wild. If you do corner one, it will behave just as the "play possum" folklore suggests: It will curl up in a ball and sometimes stay that way for hours. Scientists think that this might convince some of their predators that they are dead—and encourage them to seek other more lifelike prey. Opossums certainly use this technique if they wind up in the jaws of their enemies. They actually drop their breathing and heart rate, and they drool and give off a foul scent from their anal glands. Not surprisingly, this technique seems to discourage many of their would-be attackers.

On the "Annoying Critter Scale," the opossum is actually a fairly small pest. Although they can be a problem for chicken farmers, they don't usually cause enough grief to warrant an all-out battle. Besides, some of their behavior actually makes them fascinating animals to watch.

RABBITS

One of the reasons why classifying rabbits can be confusing is that in pet stores they are often associated with guinea pigs. Rabbits are not rodents. They don't share the same overall body structure or evolutionary history. Rabbits are actually of the order Lagomorpha, belonging to the family Leporidae. While there are sixty-three species of lagomorphs worldwide, that's really nothing when you compare it to the more than sixteen hundred species of Rodentia (the largest group of mammals in the world).

Rabbits and hares have extremely powerful and large hind legs and feet, which provide a springboard for their well-known movements. The legs of jackrabbits (members of the hare family) are much longer than rabbits, and they can actually leap up to twenty feet. In general, rabbits are more like runners, and hares are more like "bounders." Rabbits are also born underground, naked, and blind; baby hares are born aboveground, have fur, and can see at birth.

The Eastern cottontail (*Sylvilagus floridanus*) is the most common rabbit, and the one that most people have seen. They're found all across North America, east

Eastern cottontail (*Sylvilagus floridanus*)

of the Rockies. Most of the local varieties are about twelve to eighteen inches long and have short, white, puffy tails of about two inches. They like to live in forests, fields, brushy areas, and on the edges of swamps. Their ability to procreate is legendary. Each female may have as many as seven litters a year, each litter comprising up to twelve offspring. And the young that are born in the spring can actually begin to mate the following summer!

You'll find these critters burrowing in little burrows or cavities, or sometimes in holes that have been previously dug by groundhogs or other burrowers. Females construct little nests of grass and bits of their own plucked fur. This type of rabbit home is called a "form," and it's well hidden so you'll have to search to find it. A roof of grass usually covers the form, in order to keep it dry and to protect and camouflage the young. It might be tucked away in a corner of your lawn, under some bushes, or in a neighboring meadow.

Rabbits and hares have physical features that well adapt them to their environment. For example, their eyes are placed far back on the sides of their heads, so they actually have 360-degree vision. They also have large, upright ears, so they can hear well. And their sense of smell is acute, too. All of these features make it hard to sneak up on them and catch them.

Fortunately for all of us, rabbits have many enemies; otherwise the world would be overrun with these critters. Ravens and crows feed on their young. Then hawks and owls hunt them from above, while foxes, coyotes, wolves, bobcats, and other mammals hunt them on the ground. In fact, rabbits and hares are the mainstay of the bobcat's diet, and it lives in much of the same territory. Hunters also shoot close to 50 million rabbits a year across North America. Their meat is considered tasty, if not exotic, and some varieties of rabbits are trapped commercially for their fur.

In the meantime, they can be a real pest in your backyard or garden, your favorite golf course, farmers' fields, or if you own a nursery or an orchard, because they are voracious eaters. They consume grass, nuts, herbs, vegetables, flowers, and the buds, branches, and stems of many small woody plants or trees. Horticultural researchers at the University of Wisconsin have found that our cottontail friends like specific kinds of young trees, usually the ones with smooth thin bark. For example, they'll go after any of the following trees: apple, cherry, plum, and red and sugar maple. They also like young berry bushes. Once they find a good source of these kinds of foods, they'll usually stay in an area of twenty acres or less. The staff at U. of W. considers the cottontail rabbit as one of the serious pests of the state, and uses the unflattering term "vermin" to describe them.

Some of the crops that rabbits attack include clover, wheat, alfalfa, rye, beans, and lettuce. They go after these valuable food supplies with their large front teeth (incisors), which grow unchecked if not ground down by chewing. If you want to catch them in action, these foragers will be most active at sunrise and sunset. During the rest of the day they will probably be hiding in the form, where overhead predators can't see them. When a female is nursing her young, she will open the form and lie on top of the young so they can feed. Then she covers it up again to keep them safe.

If you find that some of your plants, trees, or shrubs have been "girdled" by a critter, here's how to tell if it's a vole, a deer, or a rabbit. Check the teeth marks at the edge of the area that's been damaged. You will see paired tooth marks on plants or twigs that have been neatly nipped off at a 45-degree angle. Voles will typically do more damage lower on the vegetation, and their teeth marks will be very small. Deer tend to rip or snap off the young trees or shrubs. In the winter,

CONTROLLING RABBITS

If the natural predators and the hunters aren't doing enough to curtail the rabbit population in your area, here are some additional tips to keep them under control:

- Don't encourage rabbits to move into your yard. Brush piles are an open-door invitation for them to set up a new home. Weedy debris is also tempting habitat.
- If you want to keep them away from young trees and vulnerable shrubs, you should consider sturdy fencing or hardware cloth. It should start a few inches below the soil and go up the trunk at least twenty inches.
- Live-traps are usually available from hardware stores or nurseries. The U. of W. experts recommend that you use ears of field corn, dried apples, or alfalfa for bait. During the summer months, you can also try cabbage or carrots.
- Some states allow poisons to be used for rabbits, while other regions ban their use. Thiram is a fungicide that is effective. Mothballs and blood meal are other options.
- The key is to confront your rabbit problem early. If a pair of rabbits can produce thirty-five or more offspring in a single year, this is one critter problem that you don't want to get out of control.

the shrubs or seedlings will be nipped cleanly from the snowline upward about two feet. Vole damage will likely start at the collar of the root and go upward to the snowline. Rabbit droppings are round and small, and the size of a green pea, so you'll likely find these in the same vicinity.

5

LARGER MAMMALS

It's more likely that you'll encounter some of the more sizable critters in this chapter if you live in the country, or if your property is adjacent to a park or some other wildlife habitat. And if you have a cabin, cottage, or chalet, then there's a very good chance that one or more of these larger critters has interrupted your peace and tranquility by stealing your food, or raising the water level in your lake so the dock sits four feet under the water. Once again, these animals are a pleasure to see in the wild—as long as they aren't creating a nuisance to you.

BEARS

For nearly everyone, bears evoke pleasant childhood memories. There's Smokey the bear, and Yogi, who's smarter than your average bear, and Gentle Ben. In children's literature, and in adults' hearts, the bear evokes warm emotions and fond recollections of Teddy bears. Then there's the reality of bears in the wild.

Bears are smart, just as Yogi would insist. Just as smart, for example, as the hunter who was prowling around Iron River, Wisconsin, in the fall of 1990. Stalking deer, this hunter was suddenly attacked by a black bear, who picked the hunter up and "swatted him like a badminton birdie." The hunter survived. When the local game warden was asked why the animal did this, the warden said that the bear was probably protecting its food or was simply startled. Or there might be another explanation. Perhaps the bear thought, "Today deer hunting; tomorrow bears are going to be the target." We'll never know for certain.

When bears become a nuisance to people, often they are identified and moved. This is what happened, for example, to two black bears who were roaming around the Lower Paxton Turnpike in Pennsylvania a few years ago—and it is what happens to hundreds of bears a year. Usually, in fact almost always, people are the reason that bears become a problem. Like most animals, bears prefer to avoid humans, but when given handouts, or allowed to rummage through our garbage, they often become accustomed to human habitation. Then they become dangerous.

There are three common North American bears: the black, brown, and polar bears. From here it gets a bit confusing. *Ursus americanus*, the common black bears (which are not always black) have the widest distribution, and they're found throughout Canada and along the U.S./Canada border, in wide swatches down the East and West Coasts of the United States, and in the Rockies. Brown bears (*U. arctos*) and their subspecies, the grizzlies (*U. arctos horribilis*) and Kodiak brown bears (*U. arctos middendorfii*), are found in the Northwest. Some biologists don't differentiate among the three and refer to them all as grizzlies. Polar bears (*U. maritimus*) are usually only found in the far northern reaches of the continent.

Despite their name, black bears range in color from black to reddish brown to cinnamon or gray. In the East, they are almost black, but in the West, you'll see them in their lighter colors. Some of them may also have a noticeable white blaze across their chest. They weigh from two hundred to three hundred pounds, with six hundred pounds representing the extreme. Black bears have shorter claws than grizzlies, lack a shoulder hump, and have a straight or "Roman" profile. Often black bears appear to have humps, usually when they have their heads down below their shoulders as they feed.

As the name implies, brown bears are brown, ranging in color from yellowish to dark brown. They may be three and a half feet tall at the shoulder (while on all fours) and weigh up to eight hundred pounds. Grizzlies have a hump on their backs, a muscle mass to help them dig out marmots and other ground-dwelling animals. Many grizzlies have a grizzled appearance, because their hairs are tipped with gray or they have light and dark hair interspersed in their coats. It takes time to learn to identify the grizzly. Recognizing it as "a big brown bear" won't do, because it might be a big black bear—an animal that shares the same habitat but behaves very differently.

The grizzly's average weight is three hundred to six hundred pounds, although the male, unlike the female, continues to grow throughout its life. One ten-year-old male weighed over nine hundred pounds. The grizzly can be best identified by the shoulder hump, but you will also notice its concave face, and—if you're too close—its long front claws. The Alaskan brown bear can actually get as big as seventeen hundred pounds, which makes it one of the largest land carnivores in the world.

Most bears live solitary lives, except when a mother is raising young or when a male and female pair up for a month or so in the summer to mate. Even bears at the same food source generally avoid contact with one another. Living in a loosely defined home range of about eight to ten square miles that may overlap with that of another bear, these mammals will travel throughout their area harvesting food in a season. Males have larger home ranges than females, varying in size from five to fifteen miles in diameter. Depending on the season, they use different parts of their ranges. And while they don't keep one den to return to every night, for bears that live in climates requiring it, the home range contains the den for winter hibernation.

Bears hibernate (it's actually more of a long, deep sleep) in crevices in rocks or under stumps or logs, in caves, in holes in the ground, or even under buildings.

Black bear (*Ursus americanus*)

These dens must offer shelter from the elements, so they can't be wet. On occasion, bears will hibernate in heavy brush. Not surprisingly, bears in southern regions don't go into a deep sleep in the winter months.

During the late summer and fall, bears put on weight for hibernation, building up a thick layer of fat, and then enter their dens between October and January. Before denning, bears become less active and don't eat much. In fact, a plug of material forms in the lower colon, so the bear doesn't defecate during hibernation. During hibernation, the bear's body temperature drops about forty degrees Fahrenheit, the heart rate slows, and the metabolism drops by about half. A bear may lose up to 20 percent of its body weight during hibernation. Between March and May, the bear emerges from hibernation, dislodges the fecal plug, and slowly resumes bear-like activities.

Bears are most active at dawn and dusk. They usually sleep at night, in dense brush, but when food is plentiful, they may be active all day and night. Black bears are mostly vegetarian, while grizzlies eat more meat than fruit and nuts. All bears are expert fishers but will also dig in the ground after tubers, bulbs, and roots, as well as ground squirrels. Black bears eat grasses and emerging plants in the spring, tree and shrub fruits and berries in summer, and berries, larger fruits, and nuts such as acorns and beechnuts in the fall. Both grizzlies and black bears also eat insects, especially ants, wasps, honeybees (and honey), grasshoppers, grubs, mice, voles, and carrion. Bears can be destructive feeders, ripping apart stumps or logs to get at insects or they may strip bark off evergreens to get at inner bark. All bears will eat garbage, because the food found there is easily turned into energy that their simple digestive systems can use.

In late spring to early summer, a male bear will begin to follow a female and the two will associate for almost a month as they court and mate. Other males may challenge the mated male's right to breed, and fighting may occur. After mating takes place, in late June or early July, the couple parts. Bears have delayed implantation, so although the egg is fertilized, it is not implanted in the uterus until around November. Gestation is short—six to eight weeks—and cubs born in the winter weigh only six to eight ounces. Although they're born in January or February when the mother is semi-dormant, they do receive the minimal care they need. They are blind and nearly furless, nursing and sleeping as the mother sleeps heavily. First-time mothers usually have one cub, twins are the norm thereafter.

A few weeks after birth, cubs have fur, and after forty days they open their eyes. Between March and May, depending on the temperature, the cubs will weigh about five pounds and start to emerge from the den with their mother. The cubs stay with her through the summer and into the fall, by which time they weigh about fifty-five pounds. In early winter, the mother finds a den so that she and her cubs can winter together. The next spring, the mother is ready to mate again, at which time she or her mate drives the young off. Cubs may stay together for another year, but bears become sexually mature at three and a half years, and will reproduce a few years later. Bears usually live twelve to fifteen years in the wild.

Unfortunately, the grizzly bear suffers from "a public relations problem," according to U.S. Fish and Wildlife Service officers. In 1975, this large, ferocious carnivore joined the endangered species list. Before settlers arrived in North America, the grizzly, whose range extended from northern Mexico to the Arctic Circle, numbered about one hundred thousand. Today, there are an estimated one thousand bears left. Declining habitat has been the principal reason why the grizzly is nearly gone, but predation has taken its toll, too. Hunted for fur, for sport, and out of fear, the grizzly has never been a match for a rifle.

In the United States, grizzlies make their home in Alaska, the northern Cascades region of Washington State, the Cabinet-Yaak part of Montana, the northern Continental Divide area of Montana, and, of course, Yellowstone National Park. Grizzlies are also rumored to be roaming the San Juan Mountains of Colorado and the Bitterroot Mountains of Idaho as well.

The U.S. Fish and Wildlife Service has begun a program of reintroducing the grizzly into certain parts of the country, including the San Juan Mountains. Hikers and farmers probably shouldn't expect any problems from these bears because there are only a few of them available to be relocated. Grizzlies reproduce slowly in the wild and not at all in captivity, making them tough candidates for repopulation efforts. The government's transplant plans have met resistance whenever they are announced because many people fear these bears. Indeed, grizzlies are dangerous, and they can only be reintroduced in areas where the population is scarce. Grizzlies thrive best in remote areas anyway, because these provide the seclusion they need for raising families.

While the grizzly is the one with the public image problem, the black bear is no Teddy bear either. They have adapted a little better to living near humans and therefore have held onto their traditional range better. Where their territory overlaps with that of the grizzly, the black bear usually contents itself with the lowland during the summer months (which is often where people live), because the grizzly will drive the black bear out of the higher ranges.

In *Bear Attacks: Their Causes and Avoidance*, Stephen Herrero found twenty deaths attributable to black bears between 1900 and 1980, and predation appeared to be the cause in 90 percent of the cases. By contrast, the grizzly attacks he studied over the same time period resulted in nineteen deaths, mostly when hikers surprised a grizzly, especially a mother and her cubs. To put it in perspective, you have a greater chance of being hit by lightning than being killed by a bear.

Black bears have adapted to live primarily in forested areas, while grizzlies live in spaces that are more open. The black bear, less apt to fight aggressively than the grizzly, depends on cover for survival, hiding in brush and climbing trees for protection. Still, as human populations threaten their traditional habitat, more grizzlies are moving into the forest and outlying areas.

Encounters with bears on your own turf are different than ones in the backcountry. If you can, retreat to your home or a hard-shelled vehicle if you encounter a bear outside, but keep in mind that they run faster than any Olympic sprinter, so you need a good lead. If you're sure the bear *isn't* a grizzly and it hasn't already found a food source, you may be able to bluff it away with noise and by waving your hands in the air. You'll never bluff a grizzly, however, so don't even consider it.

Bears aren't necessarily nosing around your house looking for a handout. "Bears are curious, they'll show up on your porch and look in the windows," says Heidi Youmans of the Montana State Fish, Wildlife, and Parks Department. But don't give them a reason to return: Always keep a clean and sanitary yard. If your garbage is smelly or you have any food sources outside, bears will come around for an easy meal. And once they're in the yard, they'll explore other food sources around the house. Keep garbage inside until it's time to take it to the dump or put it out for collection. Protect your gardens and orchards with an electric fence— and consider getting a big, well-trained dog, too.

Even living near a landfill or having sloppy neighbors is a cause for concern if it brings bears down around your dwelling. Once they learn about garbage, they seek it out and have an uncanny ability to find it anywhere. To discourage bears from coming around your house, keep brush cut down so they don't have cover to get close. Four walls and a roof may not be enough to stop bears that have associated the home with food; bears have torn through walls, roofs, doors, and windows in search of food.

Here's some important advice: Don't leave your children playing alone in the yard in *any* area where bears live. Fifty percent of the deaths from black bears in Herrero's study were children under age eighteen. In one case, a black bear attacked and partially ate a child who was playing near her home. If a bear is hanging around your house, you may be able to startle it away with a gas exploder, a shotgun blast, loud music, or flashing lights. In time, however, the animal will get used to the noise and lights, and the methods will become ineffective—then the animal will be more dangerous than ever. Some states and provinces regulate the shooting of bears, so check with officials before you kill a problem bear.

In the late 1980s, scientists started looking for an effective bear repellent. They hit upon hot pepper spray, a kind of mace for bears. It was tested once in 1984 by a field technician studying grizzlies. Unfortunately, the technician got too close to a bear he was tracking and the bear charged him. He sprayed the bear, which backed off. But then the bear charged again and bit the technician, who used the spray again. The bear retreated. It's now considered a good idea to keep a can of hot pepper spray at your side if you have to be away from protective shelter in bear country. You can get it at many outdoor stores.

If you live in a bear region, get information from the local wildlife agencies about the animals and what to do when you encounter them. Most of us can't interpret the bear's signals, and that makes it hard to know what to do when we see one. Your local wildlife officials are probably familiar with any problem bears and can best advise you how to handle the situation.

When you're backpacking and take along food, also take some common sense. If you don't want to risk your tent suddenly having an open door, then hang your food high and far away. *All* of your food and other provisions—even your soap, toothpaste, toothbrush, deodorant, and eating utensils. Just because the items you're carrying don't appear to have an aroma to you, that doesn't mean bears can't smell them. Anything fragrant or oily is attractive.

How high and how far should you store your food? The farther away the better. You don't want a frustrated bear to decide to inspect some other nearby human object. Besides, the farther away your food cache is, the more time you'll have to investigate that strange "noise" in the night. As for how high, you have to take into account the fact that *other* animals are interested in your food, too. Hang the food at least six feet up and out on the thinnest branch that will support its weight. Remember, animals in the woods have nothing better to do with their time than to try to get to your food.

It's a good idea to know how to recognize bear signs if you're going to be living near them and want to prevent a serious conflict. Bear signs come in many forms, some readily recognizable and others more subtle and difficult to detect. They may be on the ground, a tree, or a building. They'll provide clues and warnings of bears' presence in your area of residence.

Tracks are among the primary signs. Often there will be only partial prints, and bears will not always leave visible signs if the ground surface is dry and hard. Tracks will be five-toed, generally about four inches long and five inches wide. In the case of brown bears, the toes will be close together, the foreclaw marks twice as long as the toe pads, with tracks forming a relatively straight line. Black bear tracks have toes loosely spaced forming a curved arc, with the claw mark length not more than the toe pad length (if it's visible).

Bear feces are often found around residences, in yards, or nearby, although you may see them anywhere a bear has visited. The scats will likely be a mass of partly digested grasses, acorns, berries, insects, seeds, hair, pieces of bone, roots, wood pulp, fruits, human garbage, plastics, etc. The average sizes are roughly two and a half inches (adult grizzly bear) and one and a half inches (adult black bear) in diameter.

Buildings, trees, and other objects are often climbed, scratched, or bitten by bears as they explore areas or attempt to gain entry. This is where you'll find their claw and teeth marks. Bears also like to dig. The "plowing" of your garden, flowerbeds, or shrubbery is a sign that bears are looking for roots and vegetables; digging at the base of a building may be the sign of attempts to gain entry. Look for signs of grazing and browsing as well. Missing flower heads, or berry patches and fruit orchards that have been "picked," are signs of bear activity. (Note that these signs may also indicate deer, too.) Overturned rocks (both natural and landscape), logs, and landscape timbers will show where bears have been searching for insects.

Bears leave hair when they rub against trees, wooden and wire fences, buildings, shrubs, brush, and other objects. Hair may be found almost anywhere a bear has rubbed or closely passed. A more obvious sign will be buildings, equipment, or vehicles that have been damaged, or limbs broken off trees. In more serious cases, livestock may be missing, injured, or killed. You might even find yard furniture or toys that have been chewed and clawed.

Located near food sources and often with a good view, "day beds" are used by bears for resting. In fact, they may have several day beds throughout their range. Black bears usually sleep in trees, stretched out on large limbs, or on the

TELLTALE BEAR SIGNS

- You might find stumps, logs, or stones that have been overturned by a bear looking for grubs, ants, or other insects.
- Sometimes a tree will be literally torn apart if a bear scents a honey-filled beehive in its branches.
- Bits of dead animals might be left where a bear has been feeding. Typically, it'll leave the heads and the tails, and probably pieces of skin.
- If a bear has rummaged through a berry patch, it will be well torn up. Similarly, if it has attacked fruit trees, you'll see broken branches.
- Trees will be "scarred" where a bear has left claw or tooth prints. These are most noticeable on smooth-barked trees, such as beech or birch. A bear leaves these marks when it's scratching itself, rubbing away loose hair, or simply marking its territory.

ground in grassy areas or on conifer needles. Brown bears, however, use the ground, sleeping in grassy areas, on conifer needles, or in shallow depressions dug in soil or in snow. Day beds are often located near residential and other developed areas.

Here are some common bear problems, and strategies to outsmart these critters:

- *Rural or unoccupied homes are susceptible to break-ins.* Try installing extra security, tight-fitting doors and windows, bear-resistant fencing (chain link or electrical), or taped sounds of dogs barking or other noise.
- *Stored food and food preparation attract bears.* Keep foods in tight containers, and make sure all surfaces are free of food remnants. Barbecuing is a sure way to attract bears. Be sure to clean up and store your barbecue each evening. Don't leave the grease containers outside, attached to the barbecue. Rinse bottles and cans before you put them out for garbage or store them for recycling. Stabilize your garbage cans or Dumpsters to make them more bear (and raccoon) proof. Even the lighter fluid used for a charcoal barbecue will attract a bear. "Smoker" grills require even more precautions to keep the odors down.
- *Household garbage attracts bears.* Even "dry" trash (empty cereal boxes) has an odor, so be careful where everything is stored. Burning trash is not always a good solution. The burning smell will attract the bear, and the remnants often have food odor residue. Burying garbage is also not a good solution. Bears can easily detect it and uncover it.
- *Bird feeders attract bears.* Both the fallen seeds, and the rotting seeds, will create an odor. And suet feeders create an equally strong odor that attracts bears. Birds don't really need feeding at any time, and certainly not all year round. Consider feeding them only during the December to March bear hibernation period. Use trays to catch the spilled feed. Also try electric fences. Or keep a large dog in the yard.
- *Small pets in the yard, including rabbits and guinea pigs, and their foods will certainly attract bears.* Store all foods inside. And enjoy your pets as indoor companions.
- *Gardens and compost heaps are bear attractors.* Consider electric fencing. Keep the compost turned, and don't put in any meats or fruit scraps. Keep the garden and compost away from brush and the edge of wooded areas where the bear has cover.
- *Livestock are another big attractor, particularly pigs, goats, and sheep—as are their foods. Orchards and apiaries are also bear targets.* Use secure pens or cages, and electric fencing. Purchase feeders that minimize spillage. Good emergency lighting and/or taped noise can be used at night.

TOTAL CRITTER CONTROL

DEER

Deer are among the best known and most widely distributed of our large mammals. According to wildlife experts, the most common types of deer include *Odocoileus virginianus*, the white-tailed deer (found throughout most of North America, with the exception of some western states) and *O. hemionus*, the mule deer (found primarily in the West). In fact, the whitetail is so successful in the East that it's slowly migrating into western regions where it hasn't lived before. In addition, there are several subspecies, such as the tiny key deer (found in the Florida Keys), the smallest of North America's deer population.

With the exception of the male key deer, which weighs only about fifty pounds, adult male deer weigh about two to three hundred pounds. Does weigh from 25 to 40 percent less than bucks and are slightly less robust in their build. Deer are large creatures—about four or five feet at the shoulder—with slender legs and thick necks. Brown or tan-colored above, most deer have darker underparts. The young are well known for their white spots, which help to camouflage them when they are resting or trying to hide. Another feature of the young is that they have very little scent during the first few weeks after they're born, which helps to keep away predators.

Deer tend to live at the edge of forests or farmlands, browsing in open areas. They use the dense forest areas as cover—for shelter in winter and for escape from enemies. Deer especially thrive in agricultural areas, where fields are interspersed with woodcuts and streams. In addition to their natural foods, they thrive on our crops: corn, alfalfa, soybeans, vegetables, fruits, and grains. If farmers spot deer and alarm them, the whitetail actually raises its tail in alarm. (This may be where the term "high-tailing it" comes from.)

Deer are herbivores, feeding heavily on the bounty of summer and fall and making do in winter. In the summer and fall they eat flowers, shrubs, vines, persimmons, acorns, and cultivated plants such as corn, apples, and ornamentals. Occasionally they also eat grass. In winter, deer eat mostly twigs, bark, and the occasional evergreen. They feed most actively at dawn and dusk, when there is less danger; at midday, they are likely to be bedded down in a secluded spot, chewing their cud. At night they sleep in secure tall grass or dense brush.

Deer change their coats to match the season. At birth, fawns are rust-colored with white spots, which offer camouflage in the dappled sunlight of their woodland habitat. Adults sport a reddish-brown coat in summer, which changes to a gray-brown fall and winter coat. Another thing that changes with the season, in males, are antlers. Bucks grow them from April to August. These bony growths are then nourished by a layer of soft, vessel-filled tissue on the surface called "velvet." Different species of deer have differently shaped antlers: Those of the mule deer are forked while the whitetail's tines rise out of a central beam. Both mule deer and whitetails shed their antlers in midwinter. And it's a myth that the size of the antlers accurately tells the animal's age. Although the older bucks do tend to have larger racks, the size is a factor of the available food and how nutritious it is.

Males live in small groups of two to five most of the year, while the does live in larger groups of two to nine, including offspring. Depending on which part of the country they inhabit, deer breed between October and January. One buck may mate with a number of does; the animals do not pair off. After a gestation period of about six and a half months, does give birth to fawns, usually in May or June. While deer typically bear twins, the number of fawns relates closely to how abundant the food sources have been. Under optimum conditions, females may mate at a year old and bear twins or triplets. Some whitetails live as long as twenty years.

The deer population explosion sends them tumbling into places where they don't belong. At least that's the human outlook. In Ashland, Kentucky, a deer jumped through a one-hundred-year-old plate glass window that adorned a local art gallery. The animal ran through the gallery, smashing two artworks, and then escaped through a glass window on the other side of the gallery. Some witnesses thought dogs were chasing the deer.

A high school science class in Iowa City, Iowa, was treated to a doe running around the classroom. The school's principal said, "Just out of the clear blue it crashed through the window. I've never seen a room evacuate so rapidly. The kids were diving out of the room." And a Davenport, Iowa, radio station was interrupted by a deer that crashed through the station's window. At the time, WXLP was broadcasting a show called "Bulletin from the Boondocks."

Sometimes deer come right into your house. They're never invited, of course, they simply barge right in. In 1990, an Omaha, Nebraska, house was invaded by a buck that crashed through a window, rammed a personal computer, tore down draperies, and otherwise ruined the inside of the house. The deer had been frightened by a passing car and had run through the nearest opening—a window.

A deer ran around an Edmonton, Alberta, house for a half hour before running out the front door. The deer, which had entered the house by breaking through the screen door, "looked like an out-of-control Doberman," according to the frightened homeowner. He had spent most of his time trying to prevent the deer from running up the stairs! There's nothing you can do to discourage this type of freak mishap, but if fate sends a deer your way, just block off whichever rooms you can, open the back and front doors, and get out of the way.

And just because you live in an apartment, doesn't mean you're safe. A Bloomington, Indiana, apartment complex was invaded by five deer, one of which crashed through a glass door and window. This deer then jumped over a wall into the complex's parking lot, landing on a pickup truck. Said one waitress, "It sounded like someone was playing bumper cars in the parking lot."

Aircraft aren't immune either. A pilot crashed into a deer during a takeoff in Minneapolis, Minnesota. The private plane was demolished, and the deer was killed, but the pilot and passengers escaped unharmed. The airport manager said, "The pilot was at the wrong place at the wrong time. So was the deer."

In fact, deer will appear just about anywhere. Somehow, in 1990, a deer wandered into downtown Washington, D.C. Nobody knows how the deer got there, but it disrupted morning rush-hour traffic as it panicked among Pennsylvania Avenue cars. Exhausted and injured, it was finally darted near some dumpsters at the Old Post Office Pavilion.

Some of these unhappy encounters result when a deer wanders into our territory and panics. But in other cases, deer wreak havoc when they find us intruding on *their* territory. For example, during mating season, when a deer's hormone levels fluctuate, producing more testosterone in males, strange things are likely to happen. In Caldwell, Texas, an eight-point buck trampled a man to death. He had been walking along a rural road, looking for antique bottles, which he collected. According to investigating biologists, the deer may have seen the man bend over, picking up bottles. Apparently to a whitetail buck, being bent over, ready to attack with antlers, is an aggressive posture. And so the buck behaved as any buck would. Having spied another male in its territory during mating season, it attacked.

In another Texas incident, a buck attacked three surveyors working in the woods. One man was thrown twenty feet in the air before he landed in a creek. One Texas Parks and Wildlife officer, who was chased by a buck, described his experience this way: "It's like having an ice pick running at you at thirty-five miles per hour, with one hundred pounds of force behind it."

Clearly mating season is a time to avoid getting too close to these animals. (As is hunting season, but for a different reason—during hunting season you may be mistaken for a deer by a hunter.)

A sign of an impending attack is when a buck pushes it ears backward. A rigid body and a stiff-legged walk are two other danger signs. Wildlife experts say if a buck confronts you, don't run—walk away. There's a good chance that the deer's natural instinct to escape will take over. Of course, if the deer has begun to charge, by all means—run! Yet, attacks on humans remain rarities. Managers of parks, forests, and other natural areas say the biggest problem they face from deer comes not from the impact of deer on people, but from the impact of the burgeoning deer populations on natural resource areas.

TELLTALE DEER SIGNS

- Their tracks show cloven (split) hooves, which will be narrower than a moose or elk, with the pointed bits facing forward. Both bucks and does will drag their feet when they're browsing, so you may be able to detect this feature on the ground.
- The scat (feces) are little pellets, often oblong-shaped, that are dark and about three-quarters of an inch long.
- Because they don't have incisors on their upper jaw, deer tend to not eat very neatly. You'll notice that vegetation has been ripped rather than nipped clean.
- Deer bed down at night, so you might find a shallow little area about the size of their body. This could be in leaves, in softer brush, in snow during the winter, or in grasses near water.
- Bucks also rub the bark on young trees to mark their territory. This could show up as a "scar" on the tree, and might indicate the size and power of the buck, too.

From the human perspective there are five "kinds" of deer: the gentle and attractive ones that we might see in a park, deer that eat gardens, deer that carry ticks, deer that leap out in front of cars, and deer that hunters strap to the hoods of their cars. By far the most common and popular form of deer is the sweet and gentle animal that you spy briefly, too quickly, in the woods. After all, this is one of the most inoffensive, and cute, animals on the planet. True vegetarians and quieter than mice—how can anyone not like deer? Witness the traffic backups in Shenandoah National Park whenever deer are spotted near the road. Visitors are compelled to stop their cars and get out to snap a picture or try to feed the animal. It strolls majestically across the landscape, nibbles daintily upon a bough, and then disappears with a flash of white tail into the woods. Deer who eat our apple crops, carry ticks, or dash out in front of cars are this creature's evil siblings.

You might not like deer, however, if you have a garden. Jim Nolan, a nature writer, falls into this category. "I never consider deer to be predators, because I

White-tailed deer (*Odocoileus virginianus*)

never considered plants to be prey," he wrote in *Spiritual Ecology: A Guide to Re-connecting with Nature*. That was before he moved out of the city. Now, he says, "This gardener views his yard as a writhing hotbed where hunter and hunted are constantly devising ingenious tricks to fool one another." The basic problem, as

Nolan has discovered, is that if it's green, deer will eat it. That includes the items in your garden, as well as the tender shoots of young trees that you may have planted on your property.

Also, if it's a brown nut, a deer will probably eat it, provided it's an acorn. In fact, for each one hundred pounds of weight, a deer consumes one and a half pounds of acorns daily. That's a lot of acorns, considering that a mature buck may weigh two to three hundred pounds. So a bountiful crop of acorns can be a blessing and a curse for the gardener in deer country. A blessing, because the deer will forsake your roses, apple orchards, and other plantings for the acorns. A curse, because a large acorn crop one year means lots of deer the next year. Acorns don't fall throughout the year, so deer that eat them when they do fall will, at other times, be interested in the multiplicity of items your garden may have to offer.

Nor is the problem limited just to gardens, as farmers around Cleveland, Ohio, will testify. There, large herds of white-tailed deer have been known to consume vast amounts of wheat. One farmer described the problem this way: "It's just like a bunch of cows were turned loose." Apple orchards are also prime targets for deer. They'll eat not only the fallen apples but also the ones still in the trees on the lower branches.

At Gettysburg National Battlefield Park, maintained to resemble the farmland that it was in 1863, there rages a terrible battle—against whitetails. Bob Davidson, a management assistant at the park, says that seventeen years ago there were between four and five hundred deer on the park's four thousand acres. Today there are more than sixteen hundred deer. And what does the game commission estimate to be a suitable population? Eighty to a hundred.

The park leases land to farmers who maintain the agricultural appearance of the park, but in recent years deer damage has been so bad that farmers don't want to renew their licenses. Gettysburg has even started waiving fees if farmers can prove significant deer damage. As a result, the agricultural appearance has been maintained at great cost. An historic peach orchard was restored only by completely fencing each young tree to protect it. Farmers have apparently given up even trying to grow corn; the deer harvest it before the farmers.

And there's "political" overflow. "Right now we're very unpopular with the neighbors," says Bob Davidson. Deer are decimating landscapes and gardens all around the park. But the deer problem isn't unique to Gettysburg farmers. The Pennsylvania Farmers' Association claims $36.4 million of damage by deer every year.

It used to be that the most common reason for wanting to keep deer away was gardens. Today, a more urgent reason is Lyme disease. Named after the Connecticut town in which it was discovered, Lyme disease is a bacterial infection, second to AIDS as the fastest growing infectious disease in North America (although clearly not as serious). You can get it when you are bitten by a northern deer tick, so named because it feasts on the blood of deer, but it will also feed on pets and humans if given the opportunity. In the West, the western black-legged tick spreads Lyme disease but isn't as dependent on deer for food; it feeds on more than eighty different mammals.

Lyme disease outbreaks are most serious in eight regions across the United States: Connecticut, New York, New Jersey, Rhode Island, Delaware, Wisconsin, Maryland, and Pennsylvania. But smaller outbreaks do occur elsewhere. Fortunately, Lyme disease, if caught in its early stages, is treatable. If not caught early on, it can lead to chronic heart, tissue, and nerve disorders. (For more details, see chapter one.)

So the white-tailed deer threaten not only your azaleas but also your health. Estimates place the current deer population at 25 million—roughly the same number that roamed the country when Jamestown was a busy place. Of course by now there's much less territory for them and they're packed into small green spaces, often in populated areas.

At the Schuylkill Center, a private environmental education facility near Philadelphia, the deer population has increased twenty to twenty-five times beyond what the land can support. According to a *New York Sunday Magazine* article, the tick population at Schuylkill has increased along similar proportions.

Deer are also not popular with drivers. Car–deer accidents become common in many parts of the country during the fall and spring when deer populations are high and the animals are likely to be on the move. In some cases, they're traveling between areas looking for mates; sometimes they're looking for food. Roadside grasses are a favored fare, but deer also lick de-icing salt applied to roads in winter.

In Pennsylvania, more than forty-three thousand deer died in car crashes in 1997. Indiana had almost eleven thousand in 1998. Nationally, nearly a half-million deer are involved in these car–deer crashes, where the deer usually die and the car and the driver are injured; however, about one hundred human deaths a year occur as a result of these crashes, according to the National Highway Traffic Safety Administration. When you hit a buck and its antlers come crashing through your windshield, that's going to cause a lot of damage. And even if you don't die or get hurt in a car–deer encounter, this particular combination of metal and flesh can be both messy and costly.

You might think these accidents would occur most often on high-speed interstates or limited-access four-lane roads. They don't, because fences keep the deer off the most dangerous sections of these roads. Instead, most of the accidents apparently occur on two-lane roads. The peak times for hitting a deer are between 5 A.M. and 7 A.M. and later between 5 P.M. and 8 P.M. Dawn and dusk are the times when deer are most likely to be active. These are also the times when a driver's visibility is reduced. Many deer accidents also happen at night, when a car's headlights temporarily blind the animals. In fact, deer may freeze in the road in front of the car, or even run right at it. Trains also frequently hit deer, although there's little possibility of personal injury in those accidents.

Before you develop a fear of deer (bambiophobia?), let me add that the chances of hitting a deer with your car are slim for most people. That is, the deer accident rate per mile is small. To avoid an unwanted run-in, drive with your eyes and mind open. That means:

- Look out for deer during the times of day when they're most likely to be active (dawn and dusk).
- Look out for deer when you're driving through areas that are likely to contain deer (forest edges) or that might contain animal travel lanes (for example, a path or dirt road coming out of a forest).
- Drive more slowly when you're in deer areas. The likelihood of injury increases for drivers who go fast (and for drivers in smaller, lighter vehicles).

If you see a deer on the road, slow down as much as you can. Swerve to miss the deer, if you're sure you won't end up hitting a tree or another car. But be forewarned: When one deer has crossed a road, there's probably at least one other deer behind it.

To make roads safer, highway departments in several states are experimenting with a variety of devices. One technique that seems to work is putting reflectors along roads that deer frequently cross. When hit by a car's headlights, these reflectors create the illusion of a fence, thwarting a deer's approach. Some states have dug underpasses to give the deer a safe way to cross roads. Still others have erected off-the-road deer feeders to keep the animals away from the roads in the first place.

Another tactic is to mount a "deer whistle" on your car, but there are different opinions about whether these devices really work. The whistles, which cost about ten dollars and can be bought in catalogs and in many automobile parts stores, are installed on the car's front grille or bumper. As the car moves along the road, the whistle emits a shrill, ultrasonic cry that warns deer away, but which can't be heard by humans. The whistle can be heard up to twelve hundred feet away.

To reduce insurance claims in Indiana, one local insurance operation offered deer whistles for sale to its policyholders. The company sold fifty thousand the first year, and according to a company official, "We saw a substantial decrease in the number of deer claims." But some biologists disagree. "Deer whistles are a joke," warns one expert. Nobody's ever done a scientific study to show that they work, and nobody's studied whether deer can actually hear the whistles, he says. The only studies that have been done, he notes, are marketing studies—which turned out "okay." The California Highway Patrol takes a middle ground, claiming that deer can only hear the whistle when a car is heading directly toward the deer. Otherwise, the sound doesn't reach the deer's ears.

The deer population keeps growing, so this problem is bound to get worse. One of the reasons for the explosive growth is that the populations of many of their natural enemies, such as bobcats, mountain lions, coyotes, bears, and wolves, have dwindled. With the hunting restrictions currently in place, the whitetail is the most common—and plentiful—game animal in the eastern parts of North America.

There's no birth control for free-ranging deer at the moment, except what nature provides. A harsh winter will cut down the deer population, although the climatic changes in recent years have been unpredictable. In the late 1990s, for example, there was a string of unseasonably mild winters. And too many deer trying to forage in any given area will eventually result in the animals stripping away all the available vegetation. When there's nothing more to eat, some deer will starve and the population will decrease.

Hunting groups like to offer hunting as the "humane" alternative to starvation when populations of deer are at their peak. While there is a lot of controversy over whether hunting is, in fact, humane, safe, and ethical, there's no question that it is popular. (Ann Landers once received twenty thousand letters in response to a column she ran on illegal deer hunting.)

In spite of all the controversy surrounding it, hunting has little effect on population size; evidently, it neither stabilizes nor reduces it. The number of hunting licenses sold has actually dropped steadily since the early 1980s. In 1989 there were about a half-million fewer licenses sold than in 1981. No matter how many hunters there are, however, it would take quite an act of carnage to significantly reduce the deer population. As long as the habitat we provide is so attractive, we'll have deer right across North America.

With all the deer around, you'd think it would be pretty easy to shoot one, but it's not. Only one in four hunters bags a deer. Hunters generally oppose any effort to reduce the deer population. As a matter of fact, they often join hunt clubs so they have access to private land where the deer are treated to corn and salt licks. Since hunters traditionally go after large bucks with trophy-quality antlers, does tend to survive the hunting season and continue to reproduce.

Larger Mammals

In the absence of meaningful predators, deer will thrive. But, to survive, the animals eat everything in sight. Deer don't present a problem for mature trees, but they are deadly to young trees that are just beginning to break through the ground. By stripping an area of its vegetation, deer also deprive other creatures of food. Look for a browse line where deer have consumed every leaf, blade, needle, flower, and twig from the ground up to a height of six feet. They eat just about anything, especially when they're really hungry, and can digest more than six hundred different species of plant.

It's pretty easy to determine whether a deer is the one robbing your garden. They leave obvious signs: tracks and droppings. Look for hard dark scat, which may be the shape of an acorn or a little pellet. Without these signs, assume that the problem is a rabbit or other rodent, or a neighbor who's too lazy to grow his own vegetables. Another indicator is unevenly torn leaves; deer eat by tearing because they do not have upper incisors. Deer also feed about four to six feet above the ground, well above the reach of a rabbit.

How do you keep deer out of your garden? The only solution that's perfect is to buy all of your vegetables at the local supermarket, which is easy for a city slicker to say. Alternately, focus on the garden's perimeter. Electric fences are pretty good deer deterrents, although occasionally deer will bolt a fence. A broken wire stops the current, so you'll have to be constantly vigilant to make sure everything's working properly.

An electric fence also keeps other animals away, because they'll become conditioned to avoid the fence, too. A couple of encounters with an electric fence and deer will learn to avoid it even when it's not activated. You'll have to reelectrify the fence when new deer come into the vicinity, but you probably won't have to continually keep the fence juiced up, wasting valuable money. For smaller enclosures (forty by sixty feet), snow fencing also works well. Enclosing larger areas is usually too much for the pocketbook. Your local agricultural extension service might also be able to give you instructions on how to build a variety of deer fences.

Instead of fences, you could opt for thorny bushes. The higher the better, but even a few feet tall should work. While deer can leap over a fence as tall as seven or eight feet, once a deer miscalculates the height of a bush full of thorns, it will think twice about nibbling in your garden. Thorny trees might work well, too. Depending on what grows where you live you might select such plants as Russian olive, hawthorn, or Japanese barberry.

What follows are charts of foods that deer usually *won't* eat either because of smell or texture.

Another way to keep deer off your property is to get a dog. A small one won't do; after a couple of encounters, the deer will figure out that the dog is just a yap-

TOTAL CRITTER CONTROL

VEGETABLES, FRUITS, AND HERBS

- Chives
- Onions
- Leeks
- Garlic
- Tarragon
- Asparagus
- Fennel
- Squash
- Rhubarb
- Parsley
- Potato
- Thyme
- Broad beans
- Sage
- Rosemary
- Sweet basil
- Oregano
- Blueberry elder
- Red elderberry
- Date palms
- Serviceberry

FLOWERS

- Snapdragon
- Columbine
- Sweet William
- Hardy fuchsia
- Gladiolus
- Sunflower
- Tiger lily
- Smooth and panicle hydrangea
- Wax begonia
- Heather
- Morning glory
- Sweet pea
- Shasta daisy
- Honeysuckle
- Daffodil
- Evening primrose
- Petunia
- Chrysanthemum
- Cosmos
- Common lilac
- Marigold
- Tulip
- Poppies
- Zinnias

TREES

- White fir
- Red, silver, and sugar maples
- Birch (European white, paper, heritage, and river)
- Wormwood
- Boxwood
- Dogwood
- Juniper
- Larch
- Norway and white spruce
- Pine (Scotch, Austrian, red, and pitch)
- Oak (chestnut, white, and northern red) although deer do like the acorns
- Cherry
- Cedar
- Basswood
- Willow
- Hemlock

PREFERRED DEER TREATS

- Balsam fir
- Norway maple
- Beech
- English ivy
- Hosta
- Apples
- Atlantic white cedar
- Alfalfa
- Rhododendron
- Mountain ash
- Clovers
- Clematis
- Pumpkins
- Cranberry
- Sweet corn

per. Big dogs seem to do a pretty good job of discouraging these intruders. That said, you'll have better success if the dog is not on a leash. You want the dog to actually chase and harass the deer if this technique is going to be effective. Normally, deer won't run far when disturbed. Although they move quickly and have been clocked at 30 mph, they usually just run a short distance and look for the nearest cover. You want to be sure they aren't simply moving slightly off your property—only to return again in a few minutes.

Or, you can dissuade deer from browsing on your boughs by using one of the many deer repellents on the market. There are contact repellents that taste bad and area repellents that smell truly awful. It's best to apply the contact repellents when the plants are dormant; then animals won't feed on the buds in the winter. Naturally, you have to respray the new growth, too, and reapply them after inclement weather. You also have to remember that deer can reach up to six feet, so you have to treat high and low.

No repellent is good for large areas, but they'll work well in small gardens and yards. While repellents won't eliminate damage, they should lessen it, and the contact repellents are generally more effective than the area ones. Some of the better-known brands include: Big Game Repellent, Deer Away, Magic Circle, Hinder, Ropel, Chaperon, Gustafson 42-S, and Chew-Nott. These can be purchased at garden supply stores or through mail-order catalogs. (See the Resource Section for more details.)

Some people rely on rather grisly, smelly, old-fashioned remedies: human hair, feather meal, blood meal, or meat meal. Human hair sometimes works because deer are afraid of people and first notice humans by smell. Try human hair from a barber or hair salon stuffed into stockings or mesh bags and hung in trees or along fences. Be sure to change the hair often. Feather meal is pretty hard to come by, but you can probably get it from a poultry processing plant. Putrefied meat scraps from a slaughterhouse and blood meal are two other options. You can put this awful stuff in baggies or punctured cans and hang it in trees or around target vegetation. But be forewarned that carnivores may pull it down in the hopes of a little snack.

To keep deer from nibbling on the leaves of trees, some people even hang bars of deodorant soap from the branches. The scent of Irish Spring is supposed to be especially effective, according to deer-plagued residents around Philadelphia. Many folks swear by the old standby repellent of mothballs or flakes sprinkled on the ground or hung in mesh bags. Still, don't forget that naphthalene, the active ingredient of mothballs and flakes, is flammable. And its noxious effects aren't limited to non-humans—it affects us, too. Prolonged inhalation can bring on headaches, nausea, and vomiting. In fact, there are reports of infant deaths from skin exposure to naphthalene on blankets.

Larger Mammals

Most of the commercial contact concoctions taste awful and discourage animals from eating your sprayed plants. Unfortunately, deer have to taste it once to know how rotten it is, so there will still be some loss of leaf. The drawback: Sprays taste as bad to us as they do to animals, so you can't use any of them on food crops. Furthermore, these potions are designed to soak into the plant and render it inedible, so don't count on being able to wash it off.

Deer Away is an area repellent, also sold under the name Big Game Repellent, which is manufactured from putrescent egg solids. The product was first developed by Weyerhaeuser to protect their seedlings from foraging deer. The smell, which is mild to humans, is overwhelmingly repugnant to deer, and they won't even taste the first leaf before they decide to find dinner elsewhere. As with other sprays, you won't want to eat anything sprayed with Deer Away, which comes in powdered and liquid forms.

Anxious to protect the state's agricultural businesses, the Pennsylvania Cooperative Fish and Wildlife Research Unit conducted a study back in 1980 of fourteen repellents to find the most effective ones. The final report stated, "Our tests have shown only one repellent, Big Game Repellent, to be significantly different from no treatment at all. . . . This does not mean that the other repellents will not work in a given damage situation. We do know, however, that repellents vary in effectiveness. Some deer may be discouraged, but others may not; variable deer numbers and feeding pressure are factors to be considered."

Working with captive deer, the Pennsylvania researchers tested the "other" effective repellents mentioned above. Along with Big Game Repellent, these included meat meal, feather meal, hot sauce (about two tablespoons to a gallon of water sprayed on leaves), Chew-Nott, Flowable Fungicide, and Gustafson 42-S.

One spray repellent that was not reviewed by the Pennsylvania researchers, but which has received EPA approval for use on food crops, is called Hinder Deer and Rabbit Repellent. The spray's smell repels deer and rabbits and will work for up to eight weeks, although it's important to respray areas after three or four weeks. Hinder Deer's active ingredient is ammonium soap. It works best if sprayed as a border on the grass surrounding a protected area so the animals will learn to avoid the sprayed plants.

Niles Kinerk, director of the Gardening Research Center at Gardens Alive!, an Indiana organic product business, tests the deer repellents his firm sells. He says, "Hinder is about as effective as Deer Away!, but the nice thing about Hinder is that it has registration for use on food crops. The trick with repellents is to get one that works at a high rate, say ninety percent, and the other thing is to use it properly, get it on before the deer have begun to browse on the plants. You especially need to start spraying in the early spring when everything begins to grow." He adds that nothing, however, will be totally effective. "If you've got a starving animal and they don't have any other options, they'll still eat the sprayed plants."

Mark Fenton, of California's Peaceful Gardens, a supplier of organic products, backs him up. "No deer repellent is foolproof. The only foolproof deer repellent is an eight-foot-high fence." Fenton does, however, offer some suggestions on ways to get the most out of any repellents you use. "The thing you want to do with deer repellents is to rotate the different types frequently. You don't want to use one for two months and figure the deer is not going to get used to it." In fact, the hungrier the animal is, the more audacious it will become. They will keep testing, and eventually they will pass through the repellent and lose the fear of it," he says. "That's why, if you keep using new scents, they don't quite get the ability to keep pushing it, because they don't get used to one."

The repellent with the longest-lasting scent that Peaceful Gardens carries is Magic Circle, a bone-tar oil. Fentons warns that "It's pretty smelly . . . [and] has a strong odor." It might even keep hunters away. Another repellent he recommends is produced from a lion-urine extract, but it doesn't last as long. Neither bone-tar oil nor lion-urine extract can be used on food crops, though. Overall, you need to be relatively selective in applying area repellents. Put them where you have seen deer or where you would prefer *not* to see deer. The alternative is to have your entire yard smell ghastly.

COMMON AROMATIC HOUSEHOLD PRODUCTS THAT THWART DEER

- Castor oil
- Cayenne pepper
- Cumin
- Dry mustard
- Fabric softener sheets
- Garlic
- Mothballs
- Onions
- Rotten eggs
- Shampoo
- Tabasco sauce
- Tobacco

Some of these products can be sprayed directly onto plant leaves; others may be tied up in bags and hung from branches or fence posts. All of them must be replaced as frequently as the smell wears off. Since deer will become used to these scents, rotating repellents may increase their effectiveness.

Larger Mammals

On his Web site, the Avant Gardener suggests the following recipe to deter deer (it is also effective against slugs and insects). In a blender, put two cups of water, one to two cups of green onion tops, one or two cloves of garlic, and three eggs, with their shells. Pulverize on chop speed for two minutes. Add the mixture to a pail in which you have already dissolved a strong bar of soap. Add one to two tablespoons of any of the following: ground red pepper, hot sauce, cumin, or dry mustard. Let the mixture sit for a day or two to ripen. Apply to plants by dipping with an old paintbrush or flick the mixture on the plants. Its odor diminishes when it dries.

Here's another recipe for a mixture that many gardeners say has worked for them. In a blender, mix two eggs and one cup of water, blending well. Add one cup of skim (it must be skim) milk, and any kind of sticky substance that is organic, such as biodegradable soap. Make as many batches as you need. Put the mixture into a sprayer and spray with a light mist on vulnerable plants. This concoction has an odor that deer really detest. One taste and they're off.

Other specialists suggest that you use the hottest peppers available. (Try habañeros, which are usually available in larger supermarkets.) Combine a couple of handfuls of peppers with water in two one-gallon plastic ice cream pails, or any similar containers. Let them stand in the sun for a week. After seven days, or as soon as the stuff really begins to smell bad, strain off about two or three quarts of the liquid and use it in your pump sprayer. Refill the bucket with water and repeat the sun warming and draining procedure. On the second or third go-round, add more peppers to the pails. Don't bother throwing the old ones out, just keep adding to the mixture—the more awful it smells, the better it works. You can also add one-eighth to one-quarter cup each of the commercially available products Hinder and Palmolive dishwashing liquid to a sprayer and fill the rest with water. Spray all plants you want to protect, to the point of runoff, once a week or after a heavy rain.

Some people have hung strobe lights on fences to repel deer, although the number of lights that it would take in order to scare a single animal probably isn't worth the added charges on your electricity bill. Lights that switch on via motion detectors may help protect gardens and yards better than bright lights because they are activated only when a deer enters your premises. The sudden change in lighting will usually scare it enough so it runs away. Like scarecrows, however, lights become ineffective once deer realize that no danger follows these surprises. Perhaps the best use of a light is as a means to alert a homeowner that there is a deer—or something else even scarier—in the yard.

Dacron netting, which can usually be found in 7 x 100 foot units, can be a good deer deterrent. Drape this black plastic netting over trees and crops to keep animals away. Dacron netting can be expensive, though. Instead, try a good cloth netting, which is usually sold in six-foot widths for about a dollar a yard, from the fabric department of outlets such as Wal-Mart or other discount stores.

Invisible Mesh Barrier, a see-through mesh netting, is also available at your local garden or home improvement center or through gardening catalogs. Before using it, make sure to check with your local authorities, who may consider these mesh materials to be fences and thus regulate their height through zoning bylaws.

Sprinklers set at random intervals are a cheap and reliable method of scaring off deer. Better still are sprinklers that are motion sensitive. Heat-and-motion-activated sprinklers, such as the Sensor Controlled Animal Repeller, can be mounted in the garden and connected by a hose to your outdoor faucet. The sensors, which are powered by 9V batteries, send out a blast powerful enough to send a deer scampering, and as an added benefit help keep your lawn and garden healthy. You don't have to spend a lot of money for a ready-made system, either. You can put one together with off-the-shelf components that are available at most building supply centers.

Deer are skittish animals, easily frightened by noise or movement. A plank hung from a tree in such a way that it bangs against the trunk when the wind

CONTROLLING DEER

- Damage to plants sometimes depends on their location. If you make it harder for a deer to get to your roses, it misses one of its favorite dishes.
- No plant is totally deer resistant.
- Deer will damage your garden just as much by trampling through it as they do by eating their way through it.
- Deer browse as they travel, so any plant that they come into contact with on their daily route is likely to become a meal.
- Deer feed most heavily in the spring, when natural food may not be readily available, and in late summer, when they must fatten up for the winter. They also raid gardens more often in periods of dry weather. They're looking for plants that have high moisture levels.
- You won't be able to find anyone in agreement on what plants are deer resistant. Depending on the species of the deer and the locale of your garden, some deer-resistant plants won't work.

blows can scare them away. So, too, can wind chimes and radios that are left on. Sheets of aluminum foil or mirrors may also bother them enough to keep them away from your vegetation.

The loud detonations from gas exploders will certainly frighten away deer as well. You can set a timing device to detonate at regular intervals, but you'll need to vary the intervals from day to day so they don't get used to the noise. To keep things really interesting, you should move the machine every few days. The devices sell from about $150. As with scent repellents, deer can get used to loud noises, too. They sometimes live near airports and shooting ranges, so firecrackers and shotgun blasts may not do the job, or only deter them temporarily.

Controlling deer requires experimentation. Deer are nervous animals—you won't know what will keep them away until you try various methods.

ELK (WAPITI)

In the mountain forests and valleys of the West, and particularly in Montana, Colorado, Idaho, Wyoming (particularly near Jackson), Washington, Oregon, and northern California, you will also see the elk, or wapiti, (*Cervus elaphus*), which is a large deer with a white rump. (The name *wapiti* is a Shawnee word for "white deer," which is one of the distinguishing characteristics of this deer. The term *elk* is actually the British name for a moose, which was incorrectly applied by early explorers.) Elk stand about four to five feet tall at the shoulders, and may be as long as seven to nine feet. Their fur is reddish brown, their tails are short, and the males often have a large and spectacular rack of antlers. Females don't have antlers, and they're usually about 25 percent smaller.

In Canada, you'll see these magnificent creatures in the Rocky Mountains, and particularly in Alberta. In Banff, a popular tourist town in the mountains, elk actually walk right through the center of town. Banff is located in the middle of a national park, and the animals seem to know that they are safe and protected. A small herd of them were actually spotted on the grounds of the Banff bus station a few years ago, which became the stuff of endless jokes.

During the summer, elk might be seen grazing in groups, usually consisting of cows and calves. The bulls are frequently off by themselves. In the fall, when the bulls are rutting, you might hear their loud calls, which sound like an army bugler at work. This signals the beginning of their mating season, when they are ready to fight for control of the female harems. What follows is a dramatic battle in which the elk charge and ram each other, sometimes fighting to the death for control of a herd of females. The victor will forage and mate with the cows over the winter, then separate at the time of the calves' birth.

Elk (Wapiti) (*Cervus elaphus*)

With the exception of the park animals, elk are not always easy to spot. They move through the forest much more silently and gracefully than either moose or deer. And they also tend to feed and move at night, or in the early or late parts of the day. If you get near, they can run at speeds up to 35 mph, or they can escape by heading into the water—where they are excellent swimmers.

Because of their foraging habits, elk can be just as much of a nuisance as their deer cousins. They may also compete for grazing land with other livestock, but they pose a larger physical threat to humans because of their size and antlers. Curious tourists who try to get too close to a grazing elk are regularly taken to the hospital with various non-life-threatening injuries. It's also a mistake to get between a cow and her calf, or anywhere near a bull during the mating season. These magnificent creatures are, in fact, best enjoyed from a distance.

MOOSE

Moose (*Alces alces*) are the largest members of the Cervidae family—which also includes the white-tailed and mule deers. Interestingly, they get their name from the Algonquin word that means "cuts or trims smooth." With their flexible lips, and the top one hanging over the bottom, they are like heavy browsing machines as they work their way through the leaves, bark, and twigs from shrubs and trees.

The impressive bulls weigh between nine and fourteen hundred pounds and can be up to seven feet tall at the shoulder and seven to ten feet long. Cows are somewhat smaller at seven to eleven hundred pounds. Bulls have large, flat antlers. Their coat color ranges from tan to nearly black, and both sexes have a dewlap, a long flap of skin hanging under the chin. Their legs are long and pale, and their tails are short and stubby. Moose are found across Canada, Alaska, the Rockies, around the Great Lakes, and in New England.

Active at most times of the day, moose alternate between feeding and sleeping. They are vegetarians, and like deer, have four stomachs and chew their cud. They're fond of aquatic vegetation, so you'll often find moose wading in water to eat it. (The leaves of water lilies are one of their favorite treats.) Moose may completely submerge in search of aquatic vegetation, or simply to get away from the dreaded blackflies that share their habitat.

Moose also eat ground plants, and will push saplings down with their bodies to eat the tender tops. They prefer willows, gray and white birch, quaking aspen, balsam poplar, dogwood, cherry maple, and balsam fir. In the winter, they eat the twigs and bark of deciduous trees, conifers, and shrubs, and, of course, lose weight because their diet is so limited. Some estimates say moose lose up to half their body weight over the winter. Yet they do have one advantage for winter survival; they have extraordinarily long legs and can walk through deep snowdrifts.

Moose are large and look a bit awkward, but they can actually run up to 35 mph. They are a bit unpredictable, though, and can quickly get angry, so they should be considered dangerous at all times. They may charge and have been known to attack people, cars, horses, and even trains, but they are generally not fond of human contact and will usually withdraw before doing much harm. They are good swimmers, too; they'll even use water as a means of escape. They can

Moose (*Alces alces*)

swim about 6 mph, for a period of up to two hours. This might happen, for instance, if a moose was traveling from the mainland to an island.

Moose move around from season to season, so it's difficult to estimate the size of their home ranges. For instance, they may move from mountaintops in summer to lowlands in winter. During any one season, their home ranges are probably two to four square miles. In late fall—November to December—moose form loose groups of up to twenty members, bulls, cows and their calves, and young moose. They stay together until spring, when the group breaks apart. Through the summer, bulls are solitary or forage with younger bulls, and they remain this way until fall when they are searching for mates.

In the autumn, adult bulls are on the prowl looking for females, who are still with their previous year's calves. Cows call out when they are in estrous, and are usually in open areas. When a bull hears a cow, he moves toward her, grunting. Once they finally find one another, they stay together for one or two weeks, and then the male seeks another female and mates with her as well.

Males are very protective of their mating rights and don't allow other males to approach. If another male challenges a mated bull, there may be a fight, but first they engage in a ritual display. The bull will circle the intruder, hit shrubs with his antlers, and sway from side to side. Urinating in the dirt to create a muddy wallow is common. The bull repeatedly urinates on the ground and scrapes the area to work the liquid and earth into a wet mess. Eventually he may lie in it and roll, and the female and calf may follow suit. (Wallows may stimulate mating behavior, but even the experts aren't sure.) If this behavior does not drive off the challenger, the two may fight.

A pregnant cow remains with her yearling through the winter. She finally drives it away around June when, after an eight-month gestation, she finds a protected spot where she gives birth to her twenty-two- to thirty-five-pound calf. Moose usually have one calf unless it has been an especially good year with plentiful food, and then twins are common. In good years, when they get plenty of food and stay healthy, yearling females can mate, but they usually have only a single birth.

Calves remain hidden for two or three days. At this point, they're covered with reddish-brown, short woolly fur, and their long ears and legs contrast with their short bodies. Unlike deer, however, they have no spots. Since they're weak and gangly, the cows are very protective of their calves during this time and won't let moose or other animals get near them. Calves weigh seventy-five pounds by nine weeks old and aren't fully weaned until the cow's next calf is born the following summer. Moose hit their prime at eight to ten years, but may live as long as twenty years.

Only the bulls have antlers, which are shed each year between December and March. Growth starts in April and continues through summer. Each year the diameter of the antler base gets larger, and this means the bull becomes more dominant and able to fight his rivals to win the females each year. Bull calves have only spikes during the first year, but yearlings and two-year-olds may have spikes or branched antlers. You might see or hear a bull moose thrashing around in the bushes with its antlers. He's marking his territory and warning other bulls to keep out.

TELLTALE MOOSE SIGNS

- Like its cousin the deer, a moose browses in a ragged fashion. Because of its size and its ungainly antlers, it often leaves a mess behind as it feeds.
- The tracks of a moose will show a cloven and pointed hoof, usually about five to six inches long. Its stride in the woods is about three and a half to four and a half feet long.
- A moose will make a bed in the woods, similar to that of the deer. You'll see additional tracks and scat in the area.
- If it's been feeding on vegetation in the water, a moose's scat will be a chip or a mass. If it's been feeding on woody vegetation, it will be oblong pellets about one to two inches long.
- A moose's trail will be larger and rougher than a deer's. It's more likely to take a detour around an object as well.
- The most likely place to encounter a moose would be in a willow or aspen thicket, a swamp or marsh, or in a spruce forest. Moose are sometimes seen beside a highway when the bugs drive them out of their normal habitat.
- A rutting moose will make a tremendous bellowing sound as it charges through the woods in the fall. Think of it as an angry, approaching freight train.

Bear in mind that moose are very big animals. *So keep out of the way.* Usually the moose will try to do the same thing, which is fortunate, because moose are very big, very strong, and very determined. They also have those large antlers, which they are not afraid to use. You can wander the woods for years and not see a single moose, but when you do spot one, hold your position and let the moose make the first move. If it stands still, take the opportunity to watch or shoot pictures. If it moves toward you, interpret that message correctly and back away. While moose can run pretty fast, they sometimes have trouble maneuvering. So if an angry moose is pursuing you, zigzag around trees—better yet, climb one.

In general, when you encounter a moose, give way and get off its path. Be particularly cautious in the fall mating and spring birthing seasons when moose are really edgy. They aren't normally aggressive, but they are—as we noted earlier—unpredictable. You're most likely to see a hungry, desperate moose around your home in the winter, when it'll decimate your trees and shrubs. If you want to discourage it from feeding, try coating target trees with deer repellents such as Deer Away or Ropel. You can also try feeding it root vegetables, tree boughs, or hay. It will almost certainly move on when the weather turns warm.

Although a moose will usually stand and fight with its powerful hooves and massive rack, a weakened moose will likely fall victim to a bear or a pack of wolves—its only two real enemies. In fact, moose die more often at the hands of hunters or when hit by a car or train. Because of their size, a moose will cause a serious accident. A large moose will often slide up the hood of a car, crash through the windshield, and peel back the roof of a smaller car as if it's been attacked with a giant can opener. It's especially important to be wary of these creatures at night when you're traveling by car through densely forested areas.

BEAVERS

People know beavers as those industrious, tree-felling, dam-building, lodge-living animals. They are the animal kingdom's models of the work ethic incarnate. Nonetheless, as much as we might admire this behavior in principle, all of this industry can get to be a nuisance if it's applied to your property. Aquatic engineers with minds of their own (including their own ideas about who holds title to the pond), eager beavers ignore property lines and have no respect for your landscaping improvements.

Except for dry areas in the West and most of Florida, beavers (*Castor Canadensis*) can be found across almost all of North America. Weighing between thirty-five and fifty pounds, and reaching lengths of twenty-five to thirty inches (excluding their flat, scaly tails), these dark brown, mostly nocturnal creatures are not hard to detect in the wild. They have small eyes and ears, and are twice as large as muskrats (who don't have the unmistakable tail). Beavers are our largest rodents, but unlike most other rodents, beavers are especially adapted for a life in watery environments—living in or beside wooded lakes, ponds, streams, and other wetlands.

Beavers have an innate need to stop running water, so streamside homes are prime real estate for them. First, they'll cut down trees and build dams. Then they engineer the habitat more to their liking. They girdle trees, killing them, and leave any remaining ones to be destroyed by the flooding caused by the dam. In time, water-loving and fast-growing vegetation begins to thrive. Willow, sweet gum, poplars, and button bush spring up around the pond, creating a constant food source for the beavers.

Because it spends much of its time in the water, a beaver has warm, thick, reddish-brown fur (so luxurious that beaver pelts have been highly prized through the centuries by fur traders and fashion-makers). In addition, glands near its tail secrete an oily substance that helps "waterproof" the animal by keeping water away from its skin. Other adaptations for the watery life include small nostrils and rounded ears that automatically close when it submerges (it may stay underwater for as long as ten minutes), a transparent inner lid that slides down over the eye that allows it to see underwater, webbed hind feet, and a flat, paddle-like tail. To communicate danger, it slaps its tail against the water. And it also uses it as a prop when it's sitting upright on dry land.

Another beaver hallmark is its unusually long, sharp incisors, which have a bright orange outer covering. It uses these to girdle and cut down trees. Because the teeth wear down under this heavy labor, they grow throughout its life. These incisors are beveled on the back side to be continuously sharpened as it gnaws and chews. It works around the tree, creating a deep groove, while also depositing a telltale pile of wood chips on the ground. Trees that are about two to six inches in diameter are favorite targets, because beavers like to feed on the tender topmost branches. One beaver can fell a three-inch diameter tree in about ten minutes, a five-inch diameter tree in about a half hour.

Trees and branches are all-purpose materials to beavers. Woven together across a stream, branches are used to form dams. These vary in size according to the topography of the land. A flatland beaver may build a short (in height) but extended dam—about three feet high and a quarter-mile long. In hilly country, the dams are higher and not as long. Beavers dam up waterways to slow the speed of the water and raise its level to an appropriate height for their lodges.

Trees also provide the building material for lodges. Beavers create lodges out of large branches and logs and cover them with smaller vegetation and mud to hold it all together. They build them along the shore or in the middle of a body of water. If a beaver chooses to live beside a swiftly flowing waterway, however, it may burrow into the bank instead of building a lodge. Beavers are always improving their lodges, making them bigger and stronger. Typically, a lodge will be about four to six feet high and twenty to forty feet wide.

After the beavers finish building the mound of vegetation, they tunnel in the underwater entrances and hollow out a cavity—above the water level—for the family to live in. There are at least two and up to four underwater entrances. These effectively keep out potential predators, such as coyotes and bobcats, although river otter and mink may still swim up and attack the kits. The beaver even uses wood chips on the floor of its lodge to absorb moisture, and it creates one or more vents to let in fresh air.

You might notice mounds of mud at the water's edge of a beaver habitat; that's their way of marking their territory. The mounds are scent-marked with castor, washed out of the beaver's castor glands by urine. There may be forty or a hundred mounds in a territory, with a heavier frequency around the dam and lodge.

Rather ungainly on land, beavers don't venture out on foot more than a hundred yards or so away from their lodge to collect their food. But they may go as far as a half-mile in water, logging the far shores of a pond or river, and dragging the branches back to their lodges through the water. They swim with their webbed hind feet, while they keep their forefeet close to their chest for holding branches or pushing aside debris in the water. Flaps of skin, which fold inward and meet behind the incisors, seal off their mouths. They are useful for dragging material, too.

Beavers are especially fond of aspen, maple, alder, willow, birch, and sweet gum trees. In the West, they'll eat conifers such as the Douglas fir and pine. They don't eat the inner wood of a tree, usually only the bark, leaves, and twigs. They prefer the bark of aspen, cottonwood, balsam, poplar, apple, ash, maple, birch, alder, and willows. They'll girdle sweet gum and pines so they can eat the gum that seeps out of the trees. Beavers do like some cultivated crops and will venture away from the safety of the lodge to gather corn or soybeans. They also feed on most herbaceous and some aquatic plants. And if there are uneaten parts of the food left over, they'll incorporate them into the dams.

Beavers are generally nocturnal, active for about twelve hours a night. It's not unusual to see them during the day, however, since they begin activity in the late afternoon and go until early morning. They're most frequently sighted in the evenings, hard at work repairing dams or gathering food. If you disturb one, it can stay submerged for as long as fifteen minutes. During that time, it can also travel up to a mile to escape danger.

Beavers don't hibernate in winter and rely on food they store on the bottom of the pond, called a "cache." They swim underwater, gather some branches from the cache, and take them up into the lodge for feeding. They may also venture out in the winter to eat bark from nearby trees; however, they are very nervous on land. You may see them constantly sniffing the air for potential danger. You might also catch one using its tail to prop itself up as it sits and eats.

The lodges and dams that they build are so labor-intensive that beavers live in colonies, where they work cooperatively to build and maintain their structures and gather food. A colony usually consists of an adult male and female pair, the current season's young, and the previous year's offspring. Female adult beavers are dominant: When they speak, all colony members listen. (This might account for the fact that male beavers are expected to help care for the kits, and do so, no questions asked.) Adult females also sound the tail alarm to warn of approaching enemies.

Beavers are monogamous, breeding between January and March in the North and between November and January in the South. Gestation lasts about a hundred days, and both the male and the yearlings may be present when kits are born. The kits, two or three of them, are born with fur and are able to walk. In fact, they may go into the water after only a few days. By the time they're ten days old, they can dive and swim about, and by two months, they leave the lodge with the adults. The parents introduce tree leaves to the kits' diet early on, and after a few months the youngsters begin to feed on bark. By summer, they are pretty heavy, weighing ten to fifteen pounds. Their yearling siblings may weigh fifteen to thirty pounds by then.

As autumn rolls around, the entire family gets to work storing food for the winter and building or repairing the lodge and dam. During their second summer, the yearlings may wander away for a while, but they generally return for the winter to spend it with the family in the lodge. By their third summer, the young beavers are sexually mature and ready to start a colony of their own. They set off to find a home range and a mate, traveling the farthest migration of their twenty-year lives—possibly thirty miles.

Now, not everybody has a beaver problem, but some properties are big enough to encompass ponds and streams where beavers live. In those instances, beavers have *people* problems. Some animals clash with humans because they actively annoy us and come into our territory—they tip our garbage cans, gnaw through our walls, or nest in our attics. But the beaver is just trying to make a home in the wild, one probably not too close to human habitation.

The Fund for Animals points out several positive effects of the beaver's work:

- The dams help with flood control by holding back water and releasing it at a slow rate.
- The wetlands created by the dams provide crucial habitat for other wildlife, including migrating ducks and other waterfowl, as well as fish.
- The dams are effective for drought control, regulating a slow flow of water downstream.
- Beaver dams store water that can be used for irrigation or fighting fires.
- Backed-up water recharges the ground water supplies, including our wells.

But beavers do not live quietly; they alter the landscape when they make their home. They've caused major damage in some areas, with estimates ranging from three to five million dollars in the Southeast from flooding and crop and timber loss. (In defense of beavers, though, you have to wonder where the caretakers of the damaged land were. A beaver pond doesn't spring up overnight!) Reservoir dams have been destroyed and trains derailed because of damage done by burrowing beavers.

The average beaver-bothered homeowner has it easy by comparison—damaged trees and plants and minor flooding. Nonetheless, some homeowners aren't keen on beavers destroying their newly planted trees, especially aspens and willows. They also might not like the idea of backyard streams being dammed and flooded to form ponds and marshes.

If you want to keep beavers away from your trees, one thing you can do is fence them off with thick burlap. Wrap the material around the tree so the rotund rodent can't sink its teeth into the bark. You can also fortify the burlap with a potent chemical weapon: cayenne pepper. Spray the burlap you've wrapped around the tree with water mixed with cayenne, or spray the tree directly. Some people have also had luck with deer repellents such as Ropel and Deer Away sprayed onto the bark and foliage.

In a similar vein, some beaver defenders recommend heavy wire fences be placed six to twelve inches from a tree so that beavers can't get their teeth into the bark. The wire has to be heavy enough to stand up to thirty or forty pounds of hungry beaver pressing against it. Make sure it's firmly attached to the ground, too, so the beaver can't crawl under it. You'll also need to adjust the wires every few years to make sure you don't girdle the tree yourself. Sherri Tippie, who heads up the Colorado group Wildlife 2000, recommends concrete-reinforcing wire because it's not as visible. And Larry Manger of the U.S. Department of Agriculture in California points out that your fence should be three and a half or four feet high, because beavers will stand up to do their gnawing.

Instead of guarding trees individually, you can also install a metal fence to protect a larger area. (The fence won't keep raccoons or squirrels off the trees, but they aren't likely to cut them down, either.) Cutting is most likely to occur in mid to late fall or in the early spring. But if you try to fence in an area too close to a stream a beaver wants to dam, it might incorporate your fence into the structure.

Although they're bold about cutting down trees, beavers are skittish about unnatural objects. So, close to any tree stands that you want to protect, suspend a thirty-six inch square white flag between two poles. It will scare beavers away. Remember that beavers are nocturnal creatures; you might also want to rig up some motion-activating lighting to help discourage these critters.

Expect the greatest damage to trees close to the beavers' dam, especially to the species that are their favorite foods. Beavers prefer smallish trees such as willows, cottonwoods, and the poplar varieties, trees that regenerate new growth quickly. (Beavers actually stimulate growth in willows.) Often the trees we like aren't the first choice of chow for the beaver, but they're all that's available to them.

If a beaver does manage to fell a large tree, don't remove it immediately. Let the beaver strip it of edible material, and then turn the remains into firewood. If you take the tree away too soon, it will just have to cut down another one. And despite what years of cartoons have taught us, beavers don't plan which way a felled tree will fall. In one instance, a car was smashed as it was driving by when a beaver cut down a tree—a classic example of being in the wrong place at the wrong time. But that's the least of your worries. It could fall on a utility wire, too.

Leftovers from a felled tree will be used to block flowing water and flood an area. Beaver engineering is an inexact science. They don't figure out how much water can flow through their dams and still leave enough to protect their lodges; they stop all of the water and let it back up where it may. They can't abide a leak in the dam and will fix one immediately. That's where the problems start.

Wayne Pacelle, the national director of the Fund for Animals, understands that not everyone is willing to put up with a flooded yard or house so that a beaver family can have a nice pond. In fact, the organization offers technical assistance to folks who want to stop the flooding. Their clients even include the U.S. Army and several state governments. Beavers are animals you can live with in peace, Pacelle says. "We haven't found a case where architecture won't fix the problem." A Canadian company, D.C.P. Consulting, has even designed devices that let dams drain while not driving off the beaver.

Generally, the most effective beaver control is to allow a beaver pond to drain in a way that keeps humans and beavers happy—reduce the size of the pond but don't eliminate it. Beavers usually notice holes in their dams because they hear, see, or feel the running water. Therefore, you have to drain the water in such a way that the beaver won't sense its movement—the method must be quiet, invisible, and gentle. Simple, right? Maybe.

The Fund recommends driving a pipe (PVC or other similar material) through the dam so the water can move downstream. But, first you have to figure out the mechanics so that it will drain the perfect amount of water; if you leave the beaver high and dry, it may build a dam elsewhere, somewhere more inconvenient. Calculate the volume of water you need to remove, and install a pipe of the appropriate size. Place it well below the waterline where it won't gurgle, slurp, or burp—and be invisible, too. Making it gentle will be a little tougher. You must design the intake so the water doesn't create a current going in. One way is to block the end and perforate the sides of the intake end of the pipe. If you don't want to spend your weekends experimenting with different methods, you can simply hire someone to do the job.

Although homes rarely get flooded, roads often do. Roadside ditches and culverts are perfect dam sites—easily blocked and plenty of water to back up—so beavers often choose them. There's a wire device, called Beaver Stop, which prevents them from damming up pipes; D.C.P. Consulting sells it, and it nearly always works. You could destroy the dam, but that would only give the beavers a project. They like to build dams, so you're simply encouraging them to do what they already enjoy.

In theory, it's possible to harass the animal into moving elsewhere, but it's hard to say who will give in first—you or the beaver. Tearing out a beaver dam daily is hard on the body. Maybe they'll get the hint and look for a different place for their dam and lodge; maybe they'll stay out. If beavers are common on the watershed, however, they'll probably return to the site.

Larger Mammals

Since beavers were nearly wiped out in the last century and their populations have only recently rebounded, it's illegal to shoot or trap them in many regions. Hunters agree that it's difficult to shoot a beaver anyway, and traps hold a slow, painful death in most cases. Besides, those types of lethal solutions are only temporary. Since beavers live in groups, you would have to kill the entire family, and then it's likely that another opportunistic beaver colony will take over the site. It's better to live with the devil you know.

Live trapping is possible but the process is difficult. First, the trap itself weighs twenty-five pounds, and you will be filling that with thirty to forty pounds of beaver. Then, you're probably trapping in wetlands along the water's edge, and you have to set up the trap properly, bait it, and check it regularly. It *is* hard work, but it's worth it, says licensed beaver trapper Sherri Tippie, who had never imagined she could trap a beaver on her own. By trade, she's a hairdresser. "One day I was scrubbing the floor and watching TV," says Tippie, "when I saw a story about some beavers they were going to trap on a golf course in Aurora. I thought, 'They don't need to do that.'" She called the Rocky Mountain National Park and spoke with the ranger there, who said the park would take some beavers. Then she called the wildlife officer in Denver. The officer there was incredulous. "Do you know what it will cost to relocate those animals?" he asked. She said she'd do it, and got in touch with delighted officials in Aurora.

Next, Tippie called a trapper she knew; he told her that nobody live-traps beavers—you have to kill them. Never mind, Tippie said, she'd do it without him. She went to the Denver wildlife division and picked up two, big, heavy traps. "I got them home, sat down and started crying. I didn't know what to do." But then she did what we all should do in the same situation: She read the instructions. The first night she went out she caught two beavers.

Since then, Tippie has caught and relocated more than a hundred beavers within Colorado. For bait, Tippie uses apples, castor, and willow or cottonwood limbs woven into the end of the trap. The key to success is firmly staking the trap down. Otherwise, when a beaver swims up to take the bait that's waiting on dry land, the trap flips into the water as it closes behind it and it drowns. "The only trap I recommend is the 'Hancock' trap," she says.

What do you do with beavers? Tippie has had few problems relocating them. Federal, state, and private officials are eager beaver takers. "Beavers do more work for the environment than any other animals," she says. They do a good job restoring damaged habitats such as former cattle-grazing areas. They also slow down the erosion process in rivers. Trout fishing clubs like beavers because they create wonderful habitats for the fish. Also, the collected silt from a dam and the rotted organic material it traps make terrific soil. "Most of the beautiful meadows in the mountains are old beaver ponds," she says.

TOTAL CRITTER CONTROL

If you can catch the critter without harming it, then you have the responsibility of trying to relocate it (or the group of them) to a new and suitable territory without any other beaver tenants. The worst time to transfer animals, says the Fund for Animals, is in the late fall. The animals just won't have the energy to build a new dam and lodge as well as store food for the winter. They'll starve to death over the winter. Tippie gets around this problem by only trapping in June, July, and August, when the kits are old enough to be moved.

To reduce damage to trees over the fall and winter, she leaves tree cuttings near beaver slides so they'll leave the trees alone. She takes cuttings off the hands of lawn-care companies, saving them a trip to the landfill. Nevertheless, Tippie says, supplementary feeding isn't a good long-term solution because it's tough work feeding a bunch of hungry beavers. And don't let anyone tell you trapping a beaver isn't dangerous doings. Tippie's car has been vandalized by people who misinterpret the good intentions of her work.

Tippie sets great store in sterilization for beavers, because sometimes property owners are willing to have two but not six beavers. She's assisted a veterinarian with eight sterilizations—tubal ligations or neuterings—expensive and dangerous procedures. She thinks that soon there will be a contraceptive implant that works, too. The adult beavers then stay around to protect the dam from other beavers, but they don't have any more offspring.

If you worry about things in general, here's one reason to be concerned about beavers: They are host to many internal parasites, including *Giardia*, an organism that causes giardiasis or "beaver fever," characterized by intestinal distress and diarrhea in humans. Fecal contamination from beavers will render the area below the beaver dam dangerous for drinking and possibly for swimming, too. In the beaver's defense, the parasite is spread by other animals (domestic and wild) and humans as well. Nevertheless, if you spot beavers in your vicinity, treat the water before drinking it.

CONTROLLING BEAVERS

- Protect your trees from gnawing with spray-on products and physical barriers.
- Install devices to regulate the water flow from a beaver pond.
- Don't plant expensive trees around a beaver pond; let nature take its course and put your prized trees elsewhere.
- If you cannot live with your beavers, find a responsible way to relocate them.

6

THE HUNTERS

People living in what we call "fringe suburbs," which are suburban areas butting up against open areas, see a wider variety of wildlife—including large carnivores such as coyotes and mountain lions—than do urban dwellers. As the suburbs sprawl and encroach on the wildlife's range, these critters become more interested in nibbling on foods they ought not to eat, or on things that are not food at all. Obviously, you'll need to exercise caution when you're dealing with them. Controlling ants is not a threatening task. Facing down one of these hunters is a very different matter.

COYOTES

Coyotes, along with wolves, foxes, dogs, and jackals, belong to the family called Canidae, or canids. As this name suggests, they all have doglike characteristics: strong, athletic bodies, and powerful legs for running and leaping. The scientific name for the coyote—*Canis latrans*—actually means "barking dog." The common name—coyote—comes from Mexico's Nahuatl Indians, who called this animal *coyotl*, or "trickster."

As are dogs, these wild canids are sociable animals, but some like company more than others. Wolves live in packs, and coyotes hunt in packs or pairs, but foxes tend to be more solitary. Whether they are running in packs or hunting alone, however, most of them seem to instill some level of fear or loathing in humans. In children's cartoons, the wolf is invariably the "bad guy," and we all

TOTAL CRITTER CONTROL

know the popular tune, "Who's Afraid of the Big, Bad Wolf?" The fox may win our grudging admiration for being sly, crafty, clever, and sleek, but what about the coyote? Mark Twain may have best captured popular opinion when he described it as "a living, breathing allegory of Want. He is always hungry. He is always poor, out of luck and friendless. The meanest creatures despise him and even the fleas would desert him." Humans have long hunted, trapped, and generally despised this critter.

And there's some reasonable grounds for our dislike of them. In the summer of 1991, members of Spokane, Washington's suburban North Side were demanding that coyotes in the area be trapped and killed. Cats were disappearing, and one resident found her pet's head and shoulders about seventy-five feet from her house. Officials declined to do anything, because they couldn't use traps, shotguns, or poisons in such a densely populated area. So they waited for the naturally high mortality rate of the coyote to solve the problem and warned pet owners to keep the animals confined at night, when coyotes do most of their hunting. Residents were outraged by this hands-off approach to coyote control.

Al Manville, a senior biologist with Defenders of Wildlife, looks at the problem from a slightly different perspective than the North Side residents', though. "Which is the native species and which is the introduced one?" he asks. "The coyote was there first. Responsible pet owners take steps to prevent coyotes from getting to their pets." It's a reasonable solution. After all, leaving your pets unrestrained in coyote country is like filling up the bird feeder and encouraging the animals to come and eat. Or at least that's the position of wildlife defenders.

This argument doesn't impress many farmers and ranchers. Until 1972, when the federal government banned it, the poison Compound 1080 was used by ranchers to keep coyotes at bay. Since that time, ranchers have had to rely on mechanical traps, cyanide, and guns—methods they insist are not as effective. It's common for coyotes to be condemned as livestock killers, especially when new land is cleared for farming and the coyote shows up for a few free meals.

Meanwhile, the decimation of the wolf, the main competitor of the coyote, has opened up new habitat for *Canis latrans*. The coyote population, especially in Texas, has certainly rebounded; you can now spot them regularly around the Dallas-Fort Worth Airport. Houston has its own set of problems. Coyotes have been eating the geese and ducks that live on golf courses there.

They've made their way east as well, although they were originally western animals, and they're making a comeback in semi-rural areas, too. As we killed off all the eastern wolves, coyotes took advantage of this ecological gap and began to move toward the Atlantic, appearing in New York State sometime in the 1940s. Some biologists believe that eastern coyotes exhibit behavior that is wolflike by living and hunting in packs rather than living the solitary life of the western coyote. John Anderson, a trapper in New York, says, "They look like coyotes and respond like wolves."

Not too far from New York City, coyotes are developing a stronghold. In 1990, trappers caught more than a dozen outside of the city limits. At one point, the Westchester County Airport considered trapping a family of coyotes living near the airstrip, but then decided against it, feeling they were really harmless. Coyote numbers are rising in the county, and why not? Prey animals thrive around humans, so why wouldn't the coyote come along to take advantage of the good eats? There are plenty of rodents, rabbits, carrion, birds, bugs, and occasional deer, and pets.

Not all people dislike coyotes. Their doglike manners, mellow howl, fluffy tail, and bright eyes make them interesting, almost endearing, animals. Some gardeners actually want coyotes around, because they keep fruit- and vegetable-eating mice and rabbits in check. They keep deer at bay, too. But coyotes are not completely trustworthy in the garden; some of them seem to have developed a taste for melons and will occasionally help themselves.

Fortunately, coyotes are relatively shy and not interested in people. When on the prowl, they are hard to detect. If you were looking carefully, however, you *might* see them anywhere, in most of the United States, Canada, and Mexico. They're usually pretty small animals, with males weighing about twenty-five to thirty-five pounds and females being slightly smaller. They're mostly gray in color with a white belly, although sometimes they vary in color from red to black. Their underparts tend to be buff colored. Their legs will be a rusty red or even a bit yellowish, and if you get a good look at them you might notice a dark line running up and down their lower foreleg (although that might suggest you are *too* close).

Frequently a report of a "wolf" sighting will actually be a coyote mistaken for its cousin. The coyote is smaller, usually about twenty-five inches tall at the shoulder, and about forty to fifty inches long. Look for a bushy tail about twelve to fifteen inches long with a black tip, and remember that the coyote runs with its tail flat or between its legs while the wolf holds its tail horizontal. If you're close enough, you might also notice its nose pad. A coyote's is much smaller than a wolf's.

Wildlife experts suspect that some coyotes are actually dog-coyotes or coyote-wolf hybrids, especially in the East, which may account for the wide variation in size and color. More than a few of them have bred with German shepherds, which creates a very specific shape and coloration in the animal. Some people refer to these hybrids of coyotes and domestic dogs as "coydogs." This combination of coyote and German shepherd features can produce a formidable creature with power, intelligence, cunning, and fearlessness.

Coyotes have excellent sight and hearing, but it's their sense of smell that is especially acute. They can track down prey with their keen nose, and they can read the "signs" of their prey and their competitors in the woods. Many animals leave "scent posts" in the woods: Urine or other secretions that mark their territory or try to scare off other animals. The coyote is adept at detecting and figuring out what's happening in its territory.

In the morning or evening, you might hear a pack of coyotes, even if you can't see them. They emit a series of barks or yelps, which then might turn into a long howl ending with a series of yaps. According to experts, these vocalizations are more common in the West than in the East. Coyotes use these sounds to tell each pack member where they are, and sometimes to defend a den with pups or announce a recent kill.

Coyotes mainly enjoy rabbits and rodents such as mice and ground squirrels. But their diet also includes vegetation, fruit, carrion, frogs and snakes, and birds such as grouse and pheasant. When they are hunting in packs, coyotes can bring down much larger game. For example, they will chase a larger mammal such as a deer by using a "tag team" effort. Once the animal is exhausted, or the coyotes have lured it into an ambush, the pack will move in for the kill.

If garbage, livestock, or small pets are all that's around, they will focus on those food sources. (But in the coyote's defense, a garbage mess or what appears to be a coyote attack may often be the work of feral dogs.)

Coyotes are adaptive feeders, one reason they thrive in many areas. When they live around humans, they hunt mostly at night and during the early morning hours. In more isolated locations, they may be active around the clock.

Coyotes are adaptable to changes in the environment and can thrive in forests, deserts, mountains, grasslands, and the bush. Unlike many animals that perish when their sources of food disappear, coyotes move with their food. In fact, they've been known to travel up to four hundred miles in search of new feeding areas. They're resilient, tough, and fast. Their "cruising" speed is about 25 mph, but they can run at speeds up to forty miles an hour for short distances. They are also great leapers. Experts have seen them bound up to fourteen feet in a single leap. Another adaptive feature is their ability to swim. This allows them to chase prey into the water, or to use lakes or rivers to escape their own predators or hunters.

Locations along washes, abutting open land, or creeks are prime coyote habitats. Larry Manger, a wildlife biologist with the U.S. Department of Agriculture in California, says, "In most cases, a coyote needs a cover to work a neighborhood, but I've seen them walking down the middle of the street." They've been around long enough to become bold. If nobody is picking on them, they just continue to expand and push.

DETECTING COYOTES

- Look for them along the sides of canyons or gulches in the West. In the East, you might find them along riverbanks or in wooded or brushy areas surrounding farmlands.
- Their tracks looks very similar to dog tracks. The foreprint is about two to three inches long, and the four toes will all show claw marks.
- Their scat (which is also similar to dogs') tends to have a lot of hair in it. You'll typically see the scat on a hill or some other open area where the coyotes are gathering to watch for their prey.

Most of the year, coyote life centers upon the dens if they're raising their young, but at other times they hide in dense brush or other sheltered areas and come out to hunt. You can spot a coyote tunnel by its wide mouth (one or two feet across), which leads into another tunnel of several feet—or possibly as long as twenty-five feet. At the end of that tunnel is the actual den, which is what scientists call the "nesting chamber." The dens are secluded, maybe in a rocky area or in an underground den abandoned by a skunk, badger, or some other animal, including a fox. It could even be a culvert if the coyotes are living in a more urban environment. Once the family is disturbed, however, the female will take her young pups and move to a new location.

A coyote litter may contain as few as one pup, or it could have as many as nineteen. The breeding season is late winter or early spring, usually February to April, and the pairs may stay together for a few years or even for life. So it's not surprising that the males and females raise the young together. Before the pups are weaned at six weeks, they begin to eat meals regurgitated by the parents. Within two months, they are ready to go on short hunting trips. By the fall, the extended family usually splits apart, although sometimes they stay together until the next spring. Females are often ready to breed before their first birthday.

When coyotes take livestock, people often get worked up and demand action. John Anderson, the trapper, finds that coyotes are a lot like people—it only takes one troublemaker. When he must take action to prevent damage, he watches the pack to find the one that attacks pets or livestock, and kills it. "Make sure you don't get the dominant female," he says. "You don't need to take out five or six. You take out the troublemaker." The others will disappear, at least until the next season. Anderson is licensed by the state to do trapping and hunting of problem animals; so don't try this at home.

Keeping the coyote at bay is a little tough. At first, scare tactics such as loud radios, propane cannons, and flashing lights may keep them away. But then they get used to it. They're sneaky. If you notice an individual coyote constantly "watching" a house or "casing the neighborhood," you may have a problem animal. But Anderson says, "You can't just go out and hunt them, it won't work. They're not all bad, and most are pretty good. It's the ones who start to watch the house waiting for a meal that make the problems."

While no one seems to advocate killing all coyotes, we might want to think about behavior modification—on the part of humans. Manger says, "If you live in an area with a coyote, I recommend if you have smaller dogs, or especially cats, keep them in. My feeling about domestic animals is they shouldn't be running [loose], period." Besides, there are many other predators that might take these household pets—particularly if you live in a rural area.

If you have an enclosed yard, be sure there aren't places where the coyote can get under the fence. As a rule they won't go over, they'll go under. Many times they'll just dig themselves out a hole, about the size of a basketball. Manger says, "Just so they can get their head under and pull themselves through; they don't need much." Always monitor your fence line to be sure it's secure. But don't be surprised if an enterprising individual climbs your barrier. (Electrifying the fence will offer added protection.) Also, if you let your dogs and cats out in a questionable area, stay there with them and supervise them closely.

If your dog is large and aggressive enough, it may help to keep the coyote away from your home. But don't let that big dog roam. John Anderson reports that a male Doberman in Tenafly, New Jersey, escaped from its owner for several days one spring, bred with a female coyote, and brought her home. The female waited for the male when he went into the house to eat, and when the man walked his Doberman, the female and pups followed. The Doberman father called to the coyote with high-pitched squeaks and the female and pups came along. At the time, Anderson said, "She won't attack the man, because the Doberman accepts him." After the dog was shipped to California, the coyote went away on its own.

Standing at the door while your dog is in the yard won't necessarily prevent a coyote attack, either. Coyotes work very quickly. Manger says, "I've had people let their dogs out in the morning—two minutes and they're gone." Coyotes don't care if the pet is a thousand-dollar show dog or a much-loved cat that's been in the family for eighteen years. To them, it's just food.

Coyotes will commonly travel a long way for a meal. Manger details a coyote-killed pet case he investigated. "When the woman called me, I was a little skeptical, because I took a look at a map, and I thought, 'My God, that's a long ways from nowhere. It's all residential.' So I went by and her back fence was down. There was an easement back there where the phone lines ran, and I followed it for almost a mile. It ended up opening out into a large field, which then opened out into the river. So this coyote worked his way along this easement, looking for dogs and cats, and probably had been doing it for quite a while."

Usually when there is a coyote in the neighborhood, Anderson starts getting calls from people who are missing dogs and cats and want to know if he's trapping around the area. (If he catches any domestic animals, he either frees them on the property or transports them to the pound so their owners can come get them.) But missing pets usually indicate a coyote is around. An enterprising individual could probably live-trap a coyote, but relocation would be a problem, since coyotes seem to be living in most available habitats. And more would just move into the cleared one.

Coyotes are evidently careful planners. They'll watch a household like a burglar to get the schedule down. Then one morning when the owner lets the dog out at the same time as on the previous five days, the coyotes swoop down on the pet and kill it before the owner's eyes. When you compare coyotes to other canids, Anderson says, "They make a fox look like it's developmentally handicapped."

Some hungry coyotes hunt during the daylight, so it's never a good idea to leave your small dog confined to the backyard while you're away. Or, if you do leave your dog out, give it a "key" to the house, that is, a special doggie door that opens only when activated by a radio frequency on the animal's collar.

Coyotes are occasionally dangerous to humans. Naturally, it's the height of stupidity to put a piece of food in your hand, approach a coyote, and say "Here, doggy, doggy." Coyotes can carry rabies. Almost worse, they can bite a big piece of meat off you. Although they are aggressive animals by nature, coyotes will generally leave you alone—unless you try to feed them. Truly wild coyotes will leave people alone; the habituated ones are more dangerous. A coyote that's gotten a handout once will learn to expect a handout twice.

When coyotes start killing pets, people naturally think to protect their children. Infants are sometimes killed by wild animals, but rarely. (It seems that someone always knows about a mauling or killing done by a coyote in some *other* region.) Unattended infants and toddlers *could* be targets for a deranged coyote, so never leave them alone outdoors. Older children who are not large enough to discourage a coyote (that is, who are under forty or fifty pounds) should play close to home and come in at dusk.

Keeping coyotes away should be your first priority. If they are bothering your trash, keep it tightly sealed (a good idea in any environment, because many mammals are trash hunters). The key is to keep the cans upright; you want to

---

CAJOLING COYOTES TO AVOID YOUR YARD

- Don't let pets (or children) roam alone at night; confine animals in barns, sheds, or enclosed pens overnight.
- If you must let pets out in the evening, walk them yourself.
- Keep the brush around your house cut down close; keep the yard clean so it doesn't provide cover for rabbits, mice, or squirrels, giving a coyote a reason to hunt.
- Keep garbage secure so it's not an advertisement to feed at your house.
- Don't feed your pets outside. In other words: Don't feed coyotes.
- Traps and poisoned baits have been used to curtail coyotes; nonetheless, these critters are often smart enough to outwit their human enemies.

---

make sure that there's no way a can will overturn—the coyote's objective. Changing the aroma of the garbage can help, too, by adding mothballs, pepper, or ammonia to the can.

Coyotes will always be attracted to your dog's food. If you live in coyote country, don't feed your pets outside. Even if you bring the bowl in, there's still residue. Dogs are sloppy; they leave the scent on the porch, or spill a few crumbs. As Anderson reminds us, "When things are bad, coyotes have been found with ten pounds of cow manure in their stomachs in the middle of winter. So they'll eat anything. Even if it's just a few licks off a piece of wood."

Sometimes people get fed up with their coyote problem and turn to foot-hold traps and snares, wire mesh or electronic fences around livestock, propane exploders, sirens, and even strobe lights—such as the Electronic Guard brand. There are other alternatives, though. Well-trained guard dogs are always a good way to deter coyotes, and so are donkeys and llamas. You may see these animals grazing with sheep sometimes, and it's strictly a protective measure. Both of these large mammals are strong, stubborn, and have vicious hoofs that can kick a coyote to death. It's also important to manage your herd properly and select pastures where there is the least possibility of coyotes attacking. Experts also recommend that you quickly remove any carrion that might be in the area; it simply invites the coyotes onto your property.

The Hunters

WOLVES

The wolf is generally thought to be the smartest of all the canine creatures—and the most feared. Just think of *Little Red Riding Hood* and other similar stories. Because they travel in packs, have ferocious-looking teeth, and howl at night, we seem to have a primal fear that one day we'll be eaten or attacked by one of these fearsome creatures. In reality, there are only a few documented cases of a wolf ever attacking a human. Nevertheless, we are the wolf's only real true enemy. We hunt it for its fur, and we shoot and trap it for its destructiveness.

The common member of this family is the gray wolf (*Canis lupus*), also known as the timber wolf. (There's also a sub-species *C. rufus*, the red wolf, found in parts of Louisiana and Texas.) It's similar to the coyote, but a much larger animal: twenty-five to forty inches at the shoulder, and forty to eighty inches long. A wolf may weigh anywhere from 60 to 120 pounds, most of it pure muscle. This critter also has a larger nose pad than a coyote. As for the color, there is a huge variety among this species. You might see one that is almost black, and in the far north in winter you might see one that is almost white. Mostly you see wolves that have a kind of "grizzled" gray fur, with a long bushy tail that—like the coyote's—ends with a black tip.

Wolves used to roam all across North America, but they have been hunted and trapped extensively and their territory and habitat have been altered by civilization. As a result, you are likely to see wolves only in Alaska, across most of Canada, and in parts of Washington, Montana, and Minnesota. Their preferred habitat is forests and open tundra, whereas the coyote is making itself more and more at home in urban areas. In fact, most of the sightings of "wolves" are really coyotes, their smaller and more common relatives.

Wolves live in a highly organized society. Most of them mate for life, and the pairs then form packs of up to fifteen, although normally they're about half that size. They have a leader—the alpha male—and everyone seems to know his or her position in the pack. This is conveyed by body language; even when they're playing, there are certain submissive positions that indicate who is the boss. And all of the pack members help to raise and guard the young.

One of the most recognizable characteristics of the wolf is its howl, which can be pretty haunting on a dark, moonlit night. It uses this sound to keep the pack together, and signal where all the members are at any moment. Usually the howl of a wolf only lasts a few seconds, but other members of the pack soon pick it up—and it quickly becomes a full chorus. To some people it is a beautiful sound; to others it sends shivers down their spine. The wolves need this vocalization because they often travel great distances, and their territory may cover more than one hundred square miles.

TELLTALE WOLF SIGNS

- Their tracks will remind you of a dog's, but they'll usually be bigger—probably about four or five inches long.
- Wolves like to set up scent posts. You might see or smell these areas around rocks or stumps.
- The burrow opening to the den will be about two feet across. There might be a pile of dirt around the opening, or bits of bones from leftover kills.
- Wolf scat looks like dog feces, but you might notice lots of hair from the animals the wolf has killed.
- Wolves tend to have a local rest area where they gather. These places are often open areas in fields or woods where there are plenty of mice, one of their favorite snacks.
- You might notice lots of scat in this rest area, and perhaps even signs of food that has been buried and then dug up later.

Wolves chase and eat deer, bison, moose, caribou, and sometimes smaller mammals, and they'll eat fish, birds, berries, and insects if other larger foods are not available. They catch the larger animals by hunting in packs, using careful planning and vicious attacks. Frequently their prey will die from the repeated wounds or from sheer shock. And it's a myth that wolves chase their prey endlessly. In fact, if an animal seems to be quickly outpacing the pack, they'll give up and concentrate on an easier victim. Wolves are fast—they can run up to 30 mph—but a deer will often be able to outrun a wolf as long as the ground isn't snow-covered. A moose may also be able to stand off wolves if it is determined to fight rather than run.

There's another myth that wolves follow people on trails, particularly if they're running a dogsled team. The truth is that they're usually just using the fresh trail to find an easier path through the snow. Wolves are also very inquisitive and sometimes follow the scent of animals (and humans) out of sheer curiosity.

At the same time, wolves use their own scent, posture, and voice to determine their status within the pack. The alpha male and the dominant female are often the only wolves to breed. All the wolves then look after pups cooperatively: The

The Hunters

Wolf (*Canis lupus*)

other adults train them, protect them, and feed them. When the wolves return from hunting, for example, the pups jump and bite at the throats of the adults. This apparently stimulates them to regurgitate food for the young. These pups will grow and live to be ten to twenty years old in the wild.

TOTAL CRITTER CONTROL

Why are wolves so feared, and why have they been hunted so mercilessly? While they may have been more of a pest in the past, there's not much evidence that they're doing a lot of harm today. In Minnesota, for example, less than 1 percent of all the farmers have reported any damage that can be attributed to wolves. On the contrary, there's some evidence that the wolves actually *help* the farmers by keeping the deer population moderated. Wolves also help to weed out the sick and old mammals and keep the moose and deer populations healthy. As for our fear that they might actually attack us, there has never been a reported case of a *healthy* wolf attacking a human being.

Nonetheless, if you're having a wolf problem in your area, guard dogs, lights, sirens, fences, and a few spray repellents all offer some protection. Wolves are, in fact, pretty shy creatures, and they are relatively easily scared away. If you come face to face with a wolf and are *still* worried that it might attack, do not look directly at the wolf but don't turn your back either. This will be considered by the wolf as a non-threatening but defiant stance. For your part, all it will take will be nerves of steel.

MOUNTAIN LIONS

In the western states, mountain lions—also known as cougars, pumas, and catamounts—are coming into increasing contact with humans, both because people are moving to isolated areas and because their towns are encroaching on prime cougar habitat. As with coyotes, most complaints center on missing pets and attacks to livestock.

Mountain lions (*Puma concolor*) are considerably larger than coyotes, weighing between eighty and two hundred pounds. They are the largest cats native to North America, and range from western Canada south through the western United States and into Mexico. They are also found in the Southeast. They range in color from gray through blondish to cinnamon, and the young are yellowish with irregular spots.

People rarely see mountain lions because they're shy and mostly nocturnal. Solitary and territorial, their populations aren't very dense. The average home range is ten to twenty square miles, although they'll travel much farther than that to hunt if they have to. They often surprise their prey by attacking from above, from a tree or rock outcropping.

Mountain lions are usually born in the late winter and early spring, although they may breed at any time. A female bears one to five young and raises them by herself. Then the kittens nurse for about three months, though from the age of six weeks, the mother will supplement her milk with meat. Litters are born about

---

THWARTING MOUNTAIN LIONS

- Remove brush and tree cover that the mountain lions may use to get close to your house and outbuildings. Clear up to a quarter mile from the area around any buildings.
- Confine pets and livestock at night.
- Bright lights, loud music, or dogs may repel mountain lions, but not forever.

---

two years apart, and mountain lions will live about ten or twelve years in the wild.

Mountain lions eat different foods according to where they live, but prey on just about any animal they can kill: Everything from mice to moose, including elk, bear cubs, mountain sheep, birds, rabbits, coyote, fish, and the occasional pet. Contrary to popular myths, they don't prey heavily on livestock, and statistics from western states show that mountain lions account for less than 20 percent of the losses of livestock to wild animals.

Hunted by bounty hunters and driven out of their habitat by human populations, the mountain lion population was in danger until not too long ago. They're beginning to adjust to us—and us to them—and now the population seems to be stable, but not thriving and expanding like that of the coyote. Many states still have a mountain lion hunting season.

1

THE WINGED CRITTERS

PIGEONS

Although usually welcome, birds are sometimes inconvenient. It's not that we mind a bird's nest—we just don't want it over the front door. And some people have different levels of tolerance. A few are passionate about their birds and their bird feeders; others like particular birds (often the more colorful ones) and dislike the rest.

Most people have a poor opinion of pigeons (*Columba livia*, also known as rock doves), but you still have to marvel at these birds. In the harshest urban environments, pigeons are prolific. They are a part of Venice's charm. The Swiss Army is even giving pigeons new skills by training them as two-way messengers. Their lyrical cooing can sometimes soothe even the most ardent pigeon-hater. Pigeons are smart. They are feisty, trainable, and persistent. They are nice birds, cordial to each other, and even tolerant of cats.

Pigeons are not indigenous to North America; emigrants who ate them during their long voyages at sea first brought them here from Europe. Because a boat would occasionally arrive ahead of schedule, a handful of pigeons might escape the dinner plate. Since food was abundant in the New World, these pigeons, the earliest of which arrived in the 1600s, were allowed to fend for themselves in the wild. Almost immediately, pigeons took up residence around human settlements because these sites provided an abundance of food scraps, and because they survive only in open areas.

Pigeons can eat almost anything—grain, bread, crumbs, seeds. They thrive on handouts and garbage. In fact, handouts, offered by "nice" people and tourists, amount to as much as 75 percent of a pigeon's diet. It's not surprising to see someone sprinkle an entire bag of birdseed in a park—a favorite congregating place for pigeons. Local flocks, sometimes two hundred strong, flourish because of the generosity of these well-meaning people.

Almost three hundred years since their introduction, pigeons are rarely regarded as an important food source. Yet, these birds are still controversial. Pigeons frequently make the first page of newspapers. Here's a headline from the *Los Angeles Times* metro section, November 26, 1989: "Pigeon Trapping at Condominium in Laguna Beach Ruffles Some Feathers."

When the Laguna Sea Cliffs Apartments decided to rid itself of a flock of about sixty birds that were napping on people's balconies, the apartment owners created an uproar. The condo owners wanted to trap the birds and send them to pigeon heaven. But before the program could start, somebody discovered that it was against city laws to harm pigeons in any way. Besides, many people in Laguna Beach liked pigeons. A prominent citizen said, "We have to be more tolerant of nature and redefine what is a menace."

How do you deal with the problem of soiled balconies? One suggestion—feeding pigeons birth control pills—was quickly discarded because squirrel lovers would revolt (the pills would affect them, too). Relocating the pigeons was considered and tossed out, too, when people realized that the pigeons would "beat you driving back." This kind of stalemate is typical of pigeon problems.

Meanwhile, pigeon racing, according to the *Los Angeles Times*, "has been a little-noticed part of city living for decades." The ancient sport, however, isn't popular with all city dwellers. In Torrance, California, the community group RAP—Residents Against Pigeons—was founded to put a stop to the sport after the Torrance city council had refused to outlaw pigeon racing. Roger Mortvedt, a racer, calls these pigeons "highly bred athletes, born for racing. They don't hang out on fences and wires messing up the neighborhood, because they're trained not to do that. You can win or lose a race by seconds, so you don't want them to get in the habit of hanging around." Pigeon races are several hundred miles long and can involve fifteen thousand birds. While not thoroughbred horses, the fastest pigeons—those breaking sixty miles an hour—can command a price of one thousand dollars.

Physically, pigeons are remarkable. They are intelligent, strong, and swift. Many of their senses are superior to humans'; pigeons can hear ultrahigh frequencies, see ultraviolet light, and "sense" magnetic fields. They can sustain 60 mph flight for an entire day. And, as you may have surmised, they are fertile: at six months, a female pigeon can have six to eight baby pigeons—and have the same number every year for the next seventeen years. Pigeons are able to withstand great stress, too. Witness what happened to this British pigeon (as reported in the *Daily Telegraph*, May 20, 1989).

> A greedy pelican, apparently with a penchant for smaller feathered creatures, snatched an unsuspecting pigeon in St. James's Park, London, yesterday and tried to gobble it up. At first, things looked bad for the pigeon. It disappeared inside the pelican's pouch at one point. But a savior was in sight. A small boy among the crowd of onlookers distracted the pelican by swinging his arms and the pigeon managed to flutter free into the sunshine.

The common perception that only old, homeless people feed these birds is not actually true. Pigeon feeders, a large subset of the eighty-million-plus bird feeders around the country, are passionate about their hobby. Sometimes their concern stems from having to be so defensive about feeding pigeons; sometimes it comes simply from love for these gentle birds. Feeding pigeons grain and seed most closely approximates their natural diet, but people insist on feeding them bread, gruel, and table scraps. (If pushed to it, pigeons will eat garbage, insects, and even livestock manure.)

The problem, according to the pigeon haters, is their numbers. In the wild, that is to say in the middle of a city, pigeons breed any time of year. Whenever they feel like it, even in winter. Pigeons typically produce two eggs and nest several times a year. These messy constructions are built just about anywhere flat enough for roosting, and can take the form of crude platforms made of twigs and grass.

Both males and females care for the young, called squabs, who hatch from eggs about eighteen days after laying. The squabs feed on pigeon milk, secreted by their parents, and are weaned after about four weeks. Never ones to waste time, the parents lay new eggs before the nest is empty. Pigeons live about three or four years in the city, although it seems like they never die—they only seem to increase their population.

The truth is that most people don't like to have pigeons around. Pigeon droppings exude an acid that eats into building stone. Large amounts of the droppings kill vegetation, not to mention your nasal passages. The droppings are ugly and pigeons seem to have uncanny aim when you are wearing expensive clothing. The birds themselves carry a host of unpronounceable diseases that include tuberculosis, salmonellosis, histoplasmosis, toxoplasmosis, psittacosis, cryptococcosis, meningitis, and encephalitis. Mites, fleas, and ticks on pigeons will also nibble people and can be pests in the home.

There are plenty of reasons for not wanting one roosting nearby, but the pigeon is a steadfast, if not ferocious, foe. In their quest to coax pigeons into roosting elsewhere, people have tried all sorts of concoctions. Noisemakers, Nixalite (also called porcupine wire), low-voltage electric wire, ultrasound (which unfortunately has been known to affect pacemakers), and mothballs are often used.

But one of the most common techniques is to mount a plastic owl, because pigeons are deathly afraid of owls. Cal Saulnier of Plow and Hearth catalog says his company has sold some great horned owl replicas to the Department of Defense. Rumor has it that birds were making a mess of Blackbird secret spy planes and the owls were drafted into service to keep the birds out of the plane hangars.

These owls are familiar sights; so too is a pigeon perched on top of one. The trick is to keep the owls in motion. Experts agree that the plastic predators must be moved often, but their advice varies. Some recommend moving the replicas every few days, and others recommend moving them every few weeks. Pigeons won't take long to figure out that the stationary owl is no threat. And once the birds realize the owls can't move, let alone eat, no amount of movement is going to make them change their minds.

Some cities are taking a more ecological approach to pigeons. To reduce the flocks, peregrine falcons have been reintroduced into cities. The falcons are thriving: but so are the pigeons. In reality, falcons probably eat one pigeon (or rat) every week. The territorial falcons need a lot more space than do either pigeons or rats, and falcons have a slower reproductive rate. Falcons may never drastically reduce the pigeon population, but they do help.

Nixalite and noise are probably the best two weapons against pigeons. Nixalite is like a bunch of sharp jacks strung together, with giant barbs pointing in all directions, which prevents pigeons from alighting. Pigeons are easily startled, although they quickly return to their original place. Continual noise through radios—talk radio stations are best—or wind chimes (assuming you live in a windy place) are good tools. Firecrackers can't be beat, but you usually have to keep tossing them. If you're lucky, an occasional loud firecracker will scare pigeons away indefinitely. You have to keep up the harassment until the pigeons find new roosts. Since pigeons are usually an urban problem, devices that make continuous noise or intermittent explosions may do more harm to neighborly relations than they will aid pigeon deterrence. Gas exploders, alarms, and shotguns work, but not in the city.

Exclusion takes a little more work. Pigeons roosting in indoor areas need to be pushed out. Try mothballs or flakes or loud noises, and then block vents, eaves, and other openings with quarter-inch mesh screen, wood, glass, or even plastic netting. Mothballs can be toxic to humans, so be careful when using them around people. Just make sure you're not trapping any birds inside when you close the exits.

Netting is effective not only against pigeons, but against sparrows and starlings as well. Just about any structure that you want to keep pigeons off—buildings, ledges, detached houses, cars, even trees—can be covered with netting.

The Winged Critters

When you see netting under bridges it usually isn't there to prevent workers from falling, but to keep pigeons from roosting.

There are other techniques you can employ to prevent pigeons from hanging out. A single wire "fence" several inches high may stop pigeons from roosting on a roof or other surface. Electrifying the wire will encourage the birds to go elsewhere. A low-voltage, low-amperage current that won't harm the birds will be sufficient. Pigeons will perch on flat surfaces, but not ones at more than a sixty-degree angle. Use wire mesh, metal, or wood to create angled surfaces. In open buildings with exposed beams, use netting or mesh to block access to the upper reaches of the building.

Nontoxic repellents make roosting uncomfortable for pigeons. The tacky repellents come in a variety of forms: dry, spray, paste, and liquid. But you have to use these repellents on all surfaces in the area or the pigeons will just bunch up in a corner away from the sticky area. When you apply the repellents, make sure you don't leave any more than three inches between any strips of the substance, or the pigeons will stand there. If it's not too dusty, an application should last about a year. It may stain some surfaces, so place it on painter's tape or something you can later remove.

If you need to destroy pigeon nests that are left behind, be careful. Laurie Bingaman, on staff at the National Zoo, reminds people to take precautions against airborne diseases carried by pigeons: "Get a towel, get it damp, and tie it around your face like a bandit would, covering your mouth and nose. Then the area the pigeons have been living on, any nesting material, and any droppings should be dampened gently to avoid stirring up dust. Dispose of the mess in a plastic bag. Finally, clean the former pigeon area with a disinfectant." With the high reproductive rate of pigeons, it does little good to destroy pigeon eggs—it's an insignificant action to destroy one or two eggs when they're so easily replaced.

John Adcock, owner of a pest control company in Maryland, is a little more cautious: "The real problem with pigeons is that they're filthy animals and leave bird droppings all over. There are about twelve different lung diseases you can pick up from these critters. We use respirators—the same type of respirators you'd use if you were cleaning up hazardous waste such as asbestos."

Here's some advice from Al Geis, research director of the Wild Bird Centers of America. "If you can eliminate the habitat that attracts the nuisance animals, you've corrected the problem and you've corrected it in an enduring and fundamental way. There are many solutions that are very short-term; a lot of exterminating companies are big on short-term solutions because it creates a constant business for them. The fundamental solution is usually habitat alteration if that's possible."

Geis once worked on a bird project for the state of Pennsylvania to help eliminate a pigeon problem on a bridge over the Delaware River to New Jersey. "I ended up walking the catwalks underneath these things, and I noticed these piles of pigeon remains; there would be pigeon intestines and occasionally pigeon heads, and I asked the workman, 'What's that all about?' and he said, 'The hawks leave that behind.' Well, they have a peregrine population living in these bridges, feeding on the pigeons."

A pest control operator proposed baiting the pigeons with corn on the New Jersey side (Pennsylvania had outlawed the practice) and then substituting strychnine-treated corn for the straight corn to poison the pigeons. "My principal contribution to this project was to point out it was contrary to the provisions of the Endangered Species Act to even place these birds [peregrine falcons] in jeopardy," says Geis.

While the predators had no significant effect on the pigeon population, Geis noted that, "Poisoning the pigeons wouldn't significantly control the pigeon population as long as those nice, broad, flat surfaces in a protected situation remained." He recommended solutions that would exclude the pigeons from the bridge. For instance, pigeons can't perch on a thin wire, so if you string a thin wire about an inch or an inch and a half above the surface where the pigeons are roosting, they'll find another spot—away from you. Just make sure the wire is taut and doesn't touch the surface it is protecting.

"These animals have tremendous reproductive potential and as long as the area is attractive to them and provides suitable habitat, they are going to use it," explains Geis. Which is to say that under normal circumstances you won't be able to kill enough of the birds to significantly control the problem, but you can exclude them, or *largely* exclude them, from particular areas.

Geis suggests that a big step in outwitting critters may be changing the way humans view the problem. For example, if a sharp-shinned hawk is attacking and eating the songbirds at your feeder, perhaps the best thing to do is to let nature run its course. "It's not a real problem, it's just a wonderful opportunity to see sharp-shinned hawks. And the songbirds have great reproductive rates and high mortality to balance. If that sharp-shin doesn't eat a bird around your feeding station, he'll get one somewhere else."

Meanwhile, if you see someone feeding pigeons in your neighborhood . . . well, good luck.

STARLINGS

The problem with starlings is that they're pigs at the bird feeders. They also roost in huge numbers, fouling ledges and sidewalks. Worst of all, they get to your strawberries before you can.

The Winged Critters

Common starlings (*Sturnus vulgaris*) are about the size of a robin, and they're dark brown with light speckles all over. The biggest objection people have to the birds is that they aggressively push out native species, most notably bluebirds, flickers, purple martins, and other species that nest in cavities. They push other birds away from feeders, and then invite their starling friends to dinner. They're messy eaters and waste seed on the ground.

Starling (*Sturnus vulgaris*)

TOTAL CRITTER CONTROL

Like their fellow urban habitués, the pigeons, starlings are an introduced species, first brought to our shores in 1890 and 1891 by a nature lover who wanted all birds mentioned in Shakespeare's works to have a home in America. He didn't get them all here, but the starling is his great success. By the 1930s, the birds were spotted in Nebraska, and then in Colorado nine years later. They're so successful here that they're now found across the United States, in the southern parts of Canada, in northern Mexico, South Africa, Australia, and New Zealand. Why do starlings do so well? For one thing, they thrive in open areas like cities, towns, suburban developments, and farmland.

A big reason for their success is that they eat almost anything—seeds, fruits, insects, and garbage. Probably half of their diet comes from insects. Like pigeons, starlings are successful at reproducing themselves. They make nests just about anywhere, in buildings, tree cavities, outtake pipes, or birdhouses intended for other tenants. They also nest twice a year, laying four to seven eggs. Young starlings leave the nest after just three weeks.

Starlings roost together in huge flocks when not nesting. You're most likely to see them gathered together during the winter in dense cover out of the wind. Even though they leave the communal roost during the day, they return at night, sometimes flying thirty to forty miles round-trip. Some starlings migrate each year, while others stay put. You'll see the "winterers" perched on the top of a chimney, between trips to the local bird feeder.

The best weapons against unwanted starlings roosting are the Nixalite or Cat Claw and a sloped ledge. Wire mesh or netting installed on the underside of rafters keeps them from roosting or nesting in the upper reaches of outer buildings. Starlings find tacky repellents objectionable, so these products will keep them from roosting and nesting in an area. Try startling the starlings, too. You could use gas cartridge exploders, Mylar tape, recorded distress calls, bright lights, plastic owls, firecrackers, and drums in combination. Vary the intensity, frequency, and location of your attacks. If these methods don't work against starlings in the garden, try netting. One thing we have in common with starlings and other birds, however, is that none of us can hear ultrasound.

Starlings are more aggressive than other birds in trying to gain access to our houses. Becca Schad, owner of Wildlife Matters, a Virginia animal control company, says, "Starlings will often poke a hole in window screens with their beaks, make a hole, and then the squirrels will follow." Plain window screening won't do; it's not strong enough. If starlings are trying to build a nest, just clear everything out and cover the opening with hardware cloth. Schad won't kill baby starlings. "It can't help it if it was born in your attic. Eggs, I don't feel too bad about destroying, because you have to draw the line somewhere." And it's best to stop things early, because starlings reuse the same nest every year.

The Winged Critters

Make sure there are screens on the attic vents and the other vents in the house. Sure it's a hassle, but consider the alternative. Schad often gets springtime calls from homeowners who want her to remove young starlings from stove vents when the young ones get trapped in the duct. "When the babies get big enough to fly, instead of going outside, they fall back in the duct, and then I have to take the fan out of the stove and take the babies out that way and take them out through the kitchen."

Here's another bird story, one that really got out of hand. The staff at Wolf Trap, a national park in Vienna, Virginia, dedicated to the performance arts, was locked in a fierce battle of wills with birds for years. In 1982, when the original theater burned, the National Park Service had a chance to take control of the situation by redesigning a new theater. Chief Ranger Bill Crockett says, "In that building there were many, many nooks and crannies and ledges, and birds loved the environment. So when they built the new theater this was one of the new things that everyone asked for, that it be bird-proofed as well as possible, and the architect did try to design a building that would provide fewer little nooks and crannies. However, these birds are so inventive that even on the narrowest little thing, they can build a nest."

While they have been able to deflect the pigeons to some extent, park service staffers still haven't come up with a way to get rid of the starlings. Starlings seem to nest just about anywhere. At Wolf Trap, they have found nesting habitat in take-up reel boxes, which were designed specifically to keep birds out. The reels raise and lower the speakers over the front orchestra, and the fronts of the boxes where the cables come out are protected by a set of still brushes. Yet, the starlings squeeze in between these brushes and nest inside. "It's almost impossible!" says Crockett.

These smaller birds present as great a challenge as the larger pigeons. "Starlings go just about anywhere," says Crockett. "They are not our major problem, because their droppings are smaller. Pigeons are our B-52s. It's like a losing battle. It's an outdoor theater and as much of a natural area as we can make it, but people don't understand when they spend twenty-five dollars for a front orchestra ticket and they get a bird dropping on their three hundred-dollar dress."

Each year park service staff climb into the nether reaches of the theater to put up more Nixalite—or clear it off, because birds drop nesting materials on top of the stuff and build it up so they can sit on it. They also stuff any holes with steel wool, or block off roosting spots with hardware wire.

"Birds won't stand on the piano wire stuff," Crocket explains. "We use it over some railings. The Nixalite is actually easier to put up and maintain than the piano wire. Certainly the Nixalite is less attractive, but it's easier to rig up on some of the stuff we have to protect." Sometimes only repetitive manual labor will suffice. House lights that shine upward are protected around the edges with Nixalite, but the birds will still nest on top of the bulbs. "In the spring we have to go up there and remove the nest because when the lights go on, the nests catch on fire," says Crockett.

"The other thing we've done is to close up little holes wherever we can by filling them with steel wool or covering them entirely with a sheet of metal. The upper part of the cover spot booth, which is the booth that hangs over the balcony and contains spotlights, is completely covered with wire mesh to keep the birds from roosting on the flat surface of the booth top."

Crockett estimates that Wolf Trap's staff eliminated "about eighty percent of the problem just through those means." But he goes on to say that he is beginning to consider draping a net over the entire theater while it is closed from fall to spring. The net would discourage and stop most birds from getting in the theater, particularly in the first nesting period of early springtime. Ordinarily, birds that have two nesting periods will return to their first nesting place for their second nesting. "Once activity starts up in the theater too, around mid-May, and certainly by the end of May when performances start, you have less bird activity because of the noise and all the disruption. But birds that get in there in the early spring period and start building a nest aren't going to give up their homes, not for the loudest pop rock or whatever. They're going to settle in and get used to it."

The Park Service has tried many strategies at Wolf Trap. "We put up plastic owls; we used plastic snakes. We found they did not work. If they don't move, the birds know it. They get used to it and they realize it and know it's not real," says Crockett. "The birds were sitting on the plastic owls' heads." But in spite of everything, Crockett retains a positive outlook. "There's room to support a certain number of pigeons. The thinking is that if we could go in now and kill every bird that's in there, the population would be down a year or two and then start going up again. The best way to eliminate them is to keep them out."

SEAGULLS

Although we tend to conjure up images of beaches when we think of seagulls, these birds are actually dispersed everywhere, from coast to coast and far inland. Salt Lake City even has a monument to seagulls, the animals that saved residents from a plague of crickets by arriving at the last minute and devouring the pests. In other cities, gulls are less welcome as they dive into garbage dumpsters and plague customers at outdoor eateries.

There are seventeen species of North American gulls, but the most common include the herring gull (*Larus argentatus smithsonianus*), the laughing gull (*L. atri-*

cilla), the great black-backed gull (*L. marinus*), the California gull (*L. californicus*), and the ring-billed gull (*L. delawarensis*). They've all adapted nicely to human presence, and many species are actually increasing their numbers. The ring-billed gulls of the Great Lakes have been increasing their population about 10 percent a year since the early 1970s. Since gulls are classified as migratory birds, people can't hunt, trap, or kill them without special permits.

Seagulls are omnivorous and eat carrion, mollusks, small mammals, food from humans, and garbage. Herring gulls will actually stuff themselves until they can no longer walk, let alone fly. It's not that they're gluttons; it's an adaptive behavior from times when they went through periods of lengthy famine (an unlikely occurrence today with the availability of food in dumps and fast food parking lots!).

Seagulls are such successful scavengers that they've earned the title "flying rats" in many parts of the country. No longer content with dumps and garbage

Laughing gull (*Larus atricilla*)

cans, they're so bold as to pirate away snacks and meals right out of our hands. Many beachfront resorts in Florida protect their pool and cafe areas with an enormous stringed cat's cradle rigged above these areas. The strings zigzag wildly among the palms and poles to form a barely discernible maze. The Don Caesar resort in St. Petersburg Beach, Florida, found that this strategy reduced seagull incidents by fifty percent.

Dave Sileck, beach activity director for the Don, explains, "The only secret is to use it in a very random fashion so there is no pattern. Geometrically, it will work out to be as many angles as possible. Seagulls are pretty smart; they do eventually figure it out, so you either have to add some line or take line away every once in a while to keep them honest. They don't see the wire, they just feel it. We use a number ten-gauge fishing line, which is very thin. They feel it on their wings, they get freaked out, and they just take off. But of course if there's enough food there, they'll do anything for that." They can land outside the cat's cradle and walk under it, too.

Commercial fishponds have long used a similar arrangement, employing .4 mm steel wire strung in parallel spans eighty feet apart over the ponds. Dumps have used it, but since the attractant—rotten garbage—is so alluring, they had to place the wires closer together, at least fifteen feet. Fast food restaurants use even closer parallel spans of monofilament wire. Although the strategy does seem to work for gulls, no one knows exactly why; and the strands won't exclude pigeons or other birds. Seagulls will avoid porcupine wire and sticky bird repellents, too.

If cosmetic concerns about birds seem frivolous, consider that gulls are more likely to collide with aircraft than any other bird. Worldwide, there are some ten thousand bird strikes every year, mostly over wetlands (prime bird habitat), and 140 people have been killed from bird-plane collisions over the years. One million ducks, five million geese, a half *billion* blackbirds and starlings, and seven hundred thousand gulls fly the friendly skies along with planes over eastern North America alone. That's a lot of potential hazards in the air.

The first gull-plane crash occurred in 1912 when transcontinental pilot Cal Rodgers and his plane collided with a gull. Both were immediately killed. In 1978, a DC-10 taking off from John F. Kennedy International Airport in New York City rammed a flock of seagulls. Although the plane suffered severe damage, it was able to land and the crew and passengers were evacuated before it exploded into flames. In 1977, a private plane taking off from Chicago's Meigs Field hit a flock of seagulls. The pilot and three passengers were immediately killed, and the airport crew cleaned up 180 dead seagulls.

Aside from being three scary stories about gulls and planes, these examples point out that airports need to devise effective strategies to deal with birds. Maintenance staff must disperse the birds that want to loiter on runways, and they use many of the same tools you can buy yourself. They employ exploding devices, recordings of distress calls, flares, and exploding or whistling shells. Officials take care to eliminate food, shelter, and water that the birds might find attractive. (JFK actually drained decorative pools that attracted the birds.) And they keep up the scaring techniques at night. An airport in New Zealand even uses model airplanes in the form of hawks to scare off birds. The Royal Navy drafted falcons to keep runways clear. Some airports resorted to dead gulls mounted in contorted positions along runways. What all of them found was that the techniques worked best when combined and used diligently.

HOUSE SPARROWS

Another introduced species, the house (or English) sparrow (*Passer domesticus*), makes itself unwelcome at bird feeders and birdhouses when it bullies the other birds. House sparrows were released in 1850 in New York and have since moved from coast to coast across the United States and Canada. They're here to stay now, and we need to figure out ways to live together.

·If you have trouble telling house sparrows from our own native sparrows, look for the black bib on the male. He's the easiest to spot. He has white cheeks, a gray crown, and a tan neck. The female is a comparatively washed-out version of the male. She is streaky dull brown on her back with a dirty white underbelly.

House sparrows eat mostly plant material, but about 4 percent of their diet comes from animal food. In areas around people, they supplement their diet with highly processed foods from garbage, bread crumbs on lawns, and refuse from outdoor eateries. Baby sparrows thrive on high-protein diets; about 70 percent of their food is insects and the rest comes from plant material.

The birds can breed throughout the year, but they are most active between March and August. That's when you might see a pair of them building a nest and rearing their young. The nest is a sloppily constructed version of the other nests in the weaver family, to which the sparrow belongs, and it usually has a roof. The male selects the site and protects the territory around it, and the female incubates the four or five eggs.

They hatch in about two weeks, and the young are ready to fly after about another two weeks, although the parents may continue to feed the fledglings even after they're out of the nest. House sparrows don't travel far from their homes during the nesting period, but non-breeding adults will travel miles to a seasonal feeding area. They live about five years in the wild, and the first year is the most dangerous, with the highest mortality rate.

The greatest problem with house sparrows is that they push out other native species. There's not much you can do to keep a house sparrow away from your feeder, because feeders designed to exclude larger birds such as starlings will allow finch-sized birds like the sparrow to feed. In theory, they're fonder of a mixed feed than just sunflower seeds, and they won't eat thistle at all.

Keeping them from birdhouses is even tougher. They love martin houses, and will move in before the first martins return from their winter feeding grounds. Try blocking up the holes in the early spring so the house sparrows can't get in ahead of the martins. Your martin house should be high on a pole, away from obstructions so the martins can have space for their acrobatics. Also, you need to clean the house each season after the martins leave, or they won't come back. House sparrows can't get in openings of one and one-eighth inches or smaller, but these houses can still be used by wrens, or you can buy bluebird houses because the floor size is small.

Other suggestions for a bluebird house include making the door one and one-half inches in diameter and putting a three-and-a-half-inch screened hole in the roof. Apparently, bluebirds don't mind a summer rain, but sparrows do. Watch the birdhouses carefully to make sure sparrows don't move in. If they do, tear out the nests, daily if necessary, until they give up. Also, place the birdhouses away from human dwellings where the sparrows like to congregate—bringing us to another problem with sparrows.

You might find you have too much of a good thing with sparrows hanging around the house, roosting and building nests. You can exclude them with netting or porcupine wire, or you can try a sticky repellent. Niles Kinerk, director of the Gardens Alive! Gardening Research Center, thinks sticky repellents work best. He says, "I'll tell you what we have used, and it's extremely effective depending on what you're trying to do, and that's Tanglefoot. It's amazing. If you have roosting sparrows that you don't want, and they're making a mess, put down a string of Tanglefoot, and it's over. It works on any bird; if they get their feet in this, they don't come back.

"We had an overhang over our entrance, and it had corrugated metal cover with all these wonderful little spots where sparrows could nest, and of course they made a mess where everybody was coming into the building. So we put a string of Tanglefoot around the spot and had vengeance." The birds evidently found another place to nest, because they abandoned the spot over the door.

HERONS

When a great blue heron family nested in a pine tree that overhangs her property in Orange County, California, Jean Macnab was pleased. When they returned with friends the following season, it was too much. She cleaned up several gal-

Great Blue Heron (*Ardea herodias*)

lons of bird droppings from her patio each day, and then there were the half-eaten fish dropped from the treetops. Still, it was only when development threatened the rookery tree that someone did something about the problem. After the season's families left, a research biologist moved the empty nests to an established rookery to entice new residents to move in.

Herons are long-legged wading birds of the Ardeidae family. They also have a long neck and a sharp beak for fishing and digging in the mud of shallow waters. Herons are carnivorous, so they may regard your fishpond as their own. Centuries ago, the Japanese used a *sozu kakehi* to frighten away wild boars and other animals from the gardens. A sozu kakehi is an off-center, bamboo pipe seesaw. A gentle stream of water flows into the shorter and upper end of the pipe and fills it. The now-heavy upper end swings down, empties its water, and abruptly pivots upward, banging the lower end of the pipe on a strategically placed stone. A diaphragm between the bamboo sections prevents the water from leaking from the upper to the lower cavity. The more slowly the upper part fills, the better.

To construct a sozu kakehi you need a bamboo pipe about forty-two inches long. Glue the pipe to its dowel axle, which will turn in the holds of the two supports. To make sure that you get enough weight to tip the empty pipe downward again, you must position the axle about two inches from the middle toward the upper end of the pipe. Make sure the water supply does not touch the open pipe. Wildlife biologist Larry Manger recommends this bamboo contraption. "This one woman in particular who has one of the most beautiful koi ponds I've ever seen installed one. She has not had a heron since. Herons can be a real problem. That's something the terra-cotta pipe will give you a lot of protection on, because sometimes they'll sit there for hours, just waiting, and waiting and waiting, and when the fish does finally get in range, they're lightning fast and the fish is gone."

However you decide to deal with them, remember that herons are migratory birds, and it's illegal to hunt, capture, kill, or possess them.

CROWS

The common crow (*Corvus brachyrhynchos*) is actually quite magnificent. It's a large, blue-black bird, reaching lengths of almost two feet. And its *caw-caw* cries are known to both city and country dwellers throughout North America.

Crows are actually known for their intelligence. People often keep them as pets, enjoying their antics. Behaviorists have trained crows to solve simple puzzles; they can also mimic human speech like parrots. There is even a true story of a pet crow that was watching some young children go down a slide. Because of the way their toes are arranged, crows can't perform this simple sliding maneuver. The crow managed, however, to fetch a large bottle cap and used it as a "sled" to scoot down the metal slide. The kids were justifiably impressed.

Crows are omnivorous, eating just about anything—as shown by scientific studies that have counted more than six hundred foods in their diet. Crows are indisputably more numerous than they were before the arrival of white settlers on these shores, and in some places today they gather in roosts of more than a half-million birds. In addition to indulging a love of grains, they eat a number of harmful insects such as grasshoppers and cutworms. Insects, carrion, eggs, reptiles, fish, and other young birds make up about a third of the crow's diet. The rest comes from vegetable matter, crops, and fruits and berries. Often a family member stays high in the trees to act as a sentinel while the others feed.

Crows breed in the early spring, forming pairs to build a coarse twig and stick nest, which is lined with softer materials such as feathers, grass, and bits of cloth. Nests are usually fifteen or twenty feet high in trees, or maybe on a telephone

Common crow (*Corvus brachyrhynchos*)

pole. If trees or poles aren't available, crows may nest on the ground. Males and females share the duty to incubate four to six eggs that hatch after eighteen days. Both take care of the young. The young crows leave the nest after five weeks to feed with the partners, and they stay together as a family unit for the summer. Crows live about five or six years in the wild, although some may live as long as fourteen years.

As fall looms, the families join other families to form larger groups, and larger groups gather to form immense congregations, so that by late winter, in some areas, the crows may number in the millions. They disperse during the day to feed, but return at night to be with their kindred. As a result, you might need to use noisemakers and visual frightening devices to discourage crows from holding their winter reunions too close to your house. Gas exploders, recordings of distress calls, Mylar ribbon or balloons, alarms, or bright lights may scare them away. Vary the volume and location of the surprises so the birds never know quite what to expect.

Most pest control experts do not think of crows as a problem for homeowners. Becca Schad says, "I've had calls from people complaining about crows eating from their bird feeders, but it ended up that the people were putting out corn [which crows love]. And I thought, 'Well, you're feeding the birds . . . and you're getting results.'" So, if crows bother you at your feeders, try setting up a separate feeding area for them and spread cracked corn for them to eat. If they bother your garden, scattering corn in a selected area may also lure them away from your seedlings.

WOODPECKERS

The woodpecker is well adapted to its role as the jackhammer of the animal world. Its hard, sharp beak can break holes in wood so it can reach its long tongue inside and grab the insects. A woodpecker can balance easily on the side of a tree because it has two toes facing backwards and a stiff tail for support.

Twenty-two species of woodpecker (which make up the family Picidae) can be found in North America, and you'll usually see them in wooded areas, where they find food and shelter. Woodpeckers have adapted somewhat to development and find good eats and a place to live in telephone poles, fences, and other wooden structures. In addition to boring insects found in wood, some woodpeckers eat insects found on the ground, such as ants, as well as berries, nuts, fruit, and some seeds. The sapsucker also eats sap, as its name implies, in addition to insects.

Betsy Webb, curator of urban wildlife at the Denver Museum of Natural History, has lots of experience with woodpeckers. "In the spring there is a particular problem with flickers, a type of woodpecker. They drum on the cedar siding." During the spring breeding season, the male flicker will hammer on wooden siding, and it reverberates loudly. This both establishes a territory and advertises the male's availability. The louder they drum, and the more the wood reverberates, the better able they are to set up a territory and attract mates. Sometimes they'll actually excavate a nest hole in the side of a house, often in cedar since it is soft. "In doing that, they sometimes cause damage to houses up in the thousands of dollars. They create a series of holes and the whole siding will have to be replaced, and that's very expensive," says Webb.

BUILDING A FLAPPER

Try this solution for dealing with these birds:

- Get two one-square-foot pieces of wood, matching the type of siding on the house.
- Nail them together at one corner so that they flap together.
- Secure the back piece to the wall where the damage occurred.

The flicker will focus its attention on the flapping piece of wood, which reverberates even more loudly. Set up a series of them to try to concentrate the flicker damage on the squares. Just remove them after the breeding season.

But what about the noise? Webb explains, "When you build a house in deer or flicker territory, you are going to be living with those species, and you can either have a cycle of negative problems with them year after year, or you can put in place very creative solutions to live in harmony with them." If woodpeckers are beating on a metal chimney, try wrapping it in burlap to muffle the sound. (Take down the burlap, however, before you start a fire!) They might get frustrated and find another boomer. Or try a lure—put up a metal tin some distance away from the house.

You can also hang high-quality images of birds of prey. Use flat silhouettes, such as hawks, or the plastic owls mentioned earlier. They need to be a thick reproduction of a substantial, threatening predator. If it's placed before the breeding season starts, it is likely to warn off a bird. If you put if up after the season has already started, it is not as likely to change the behavior. Birds aren't stupid; they're well adapted to their environment and are extremely opportunistic. Anything you can do to fool them must be done well. Move the silhouettes or predator reproductions occasionally so the flicker doesn't get used to them. Try banging every morning with a pot and pan—but you'll have to do it every morning at five!

You can also attract the flickers to trees in your yard (and away from your house) by putting up a nesting box in your yard. Your local Audubon chapter will have good suggestions for the specifications for flickers or for your local woodpecker population. Birds have specific nesting requirements; they'll select a box that fits their specifications for entrance hole, depth, dimensions, and placement on the tree. These species are particular and just any nesting box won't suffice.

In many parts of the country, there's a tight housing market for the birds. Mark Westall, a naturalist on Sanibel Island, Florida, says, "People build big wooden houses, clear away all the vegetation, take out all the palm trees, and put in a sod lawn, then they can't understand why the woodpeckers are poking away on the side of their house.

"Lots of times a woodpecker will poke a big hole in the side of the house because it's looking for a nesting cavity, and what you can do is put a nest box right over the hole where it's working." If you give the woodpecker what it wants, you'll save your house. Nesting woodpeckers will usually chase away other woodpeckers, but you may have to put a nest box up on both sides of your house, because one pair may nest on one side and another on the other side. The wood siding survives and you get to enjoy watching the woodpeckers raise their young.

Also, don't be so quick to remove dead trees from your yard; they will give the woodpeckers a place to live and maybe feed. It works for Westall: "We don't have a lawn, we have pruning shears. We prune the jungle away from the side of the house and we can watch the pileated woodpeckers working on trees ten feet away from the side of the house. We don't have any problems with them working on the house because we still have plenty of habitat here for them." Friends of his even acquired a couple of dead cabbage palms to put on their property so they could attract more woodpeckers.

Not everyone is so enthusiastic about woodpecker neighbors, but frustrated homeowners should be careful before they do anything drastic to rid themselves of these birds. Woodpeckers are classified as migratory songbirds, and it is illegal to harm them in any way. Also, if your house is wood, it's possible the woodpecker is doing you a favor; your home could have a termite or bug infestation you don't know about.

Keep in mind that often woodpeckers are interested in getting at the bugs in your house, not in making little holes in your house per se. Thus, getting rid of the insects that may have burrowed into the side of your house is one of the most effective steps you can take to deal with woodpeckers.

The Winged Critters

Westall says, "They make long strips up and down where they poke away the wood—usually on the corners of the house where the paneling comes together, and builders put trim over it to cover it up. Ants like to go up into your attic and follow that trail underneath that trim." The woodpeckers know that the food is there and they start cutting away at those strips. If you want to stop woodpecker damage, you have to stop the ants, and solutions range from boric acid in their nests to a professional exterminator to mothballs in the attic.

Westall's advice is applicable to dealing with all animals, including woodpeckers: "The main thing is to think like wildlife. Say to yourself, 'That woodpecker isn't damaging my house because he doesn't like me; he's doing it because that house gives it something it's looking for. Now am I smart enough to figure out how to give that to the bird without causing damage to my house?'"

If you have a garden you will have noticed that you have created one very large bird feeder. That's okay, if that was your intention, but most people don't envision this as the main purpose behind their garden. Birds have an advantage over most other animals: they can fly. That means fences are extraordinarily useless. But bird netting is not. A large garden may make netting difficult, while a smallish garden can support it very well. Netting will reduce bird infestation nearly perfectly.

Mark Fenton, president of Peaceful Gardens in California, offers this recommendation: "Row covers are good bird and insect protection. The barrier method works. There are some bird repellents, the Savanna bird repellent, for example; or you can make one at home with Vaseline and cayenne pepper. I think the scare-eye balloons are one of the more effective in bird control. I've used them personally. The birds get pretty skittish; it won't stop them completely, but they fly in and peck something, see that thing and fly out again. But they won't stay. The main thing if you have a bird problem . . . is to use various different bird things all the time. In my own orchard, I put up some scare eyes and Mylar tape this year and have way, way better control than I had last year with nothing there. Last year they got all my apples before they ripened."

While large nets are effective in keeping birds away from fruit, they are sometimes impractical. Al Geis, research director of the Wild Bird Centers of America, suggests that often an alternative food source will distract birds away from your crops. "A food they like just as well that fruits about the same time is tartarian honeysuckle, an exotic plant. Most of our native and even our ornamental fruit-bearing plants fruit too late. The fruit doesn't become available when strawberries or blueberries are available." While the tartarian honeysuckle does not spread aggressively like other exotics (the notorious multiflora rose, kudzu, and Japanese honeysuckle), you may want to check with your local agricultural office before you plant it.

An exploder could be equally effective against small intruders. Fenton explains that birds won't grow accustomed to the sound "because they never really stay around. The birds just don't live right next to it. They'll fly in and get scared away." The noisemakers should be placed in the middle of fields or orchards, far away from civilization, and allowed to boom all day long. The device doesn't shoot out flames, so it can be left unattended.

Of course, bird problems don't stop at the bird feeder or garden. They can come right into your living room, so don't underestimate the importance of excluding birds from your home using chimney cages and screening over vents. A Los Angeles family was stunned when hundreds of migrating Vaux's swifts swooped down their chimney and into the house. The swifts, on migration from Central America to the Pacific Northwest, became confused when they mistook the chimney for a sheltered nesting place and then ended up in a strange living room. They spread soot and droppings throughout the house before exiting through open windows and doors. Dozens of the birds died in the attempt.

CANADA GEESE

Although the familiar sight of Canada geese flying in a V-shaped formation is a lovely sight in the sky, these large birds are becoming increasing nuisances on the ground. They're the most widespread geese in North America, and that's the problem—there are just too many of them. Canada geese are now so prevalent that there are eleven different subgroups of this bird, each one found in a particular area of North America. In some smaller regions, the native population is less than one thousand birds, but for some of the more common varieties spread over large areas, there may be more than a million birds. Because geese are so adaptable and now live comfortably in urban and rural environments, it's not surprising that they are listed as one of the major pests of the feathered variety.

Branta canadensis is a relatively large, plump bird that may weigh anywhere from three to fifteen pounds and measure from twenty-two to forty-five inches long. The smaller varieties are found in the tundra of the high Arctic. These are often referred to as "cackling geese" because they have a more high-pitched call than the distinctive honking of their larger kin (often called "honkers"), found in the warmer southern regions.

You can identify Canada geese by their stout bodies with light brown wings and light tan or cream-colored breast feathers. They have white feathers under their tails and black heads and necks. One of their more distinctive features is their "chin strap"—white feathers that wrap around the lower part of their necks. Their prominent bills will be brown or even black, as are their legs and large webbed feet. Both males and females have similar coloring, although males are generally bigger than females.

Females will breed within three years, usually mating for life with one gander, and either sex may grieve to death if it loses its mate. Goose eggs are incubated in a very simple, hollow, bowl-shaped nest that the females scrape out of the ground and then line with goose down, grasses, and other available vegetation. Usually these nests will be within a thousand feet of a lake, river, or pond, which will provide water for drinking, cleaning, and preening. Water also offers geese an avenue of escape when predators approach.

The size of their "clutch"—eggs laid at one time—may range from two to twelve, but usually there will be about five or six dull white eggs. These will appear in the nest in the spring (earlier in the warmer southern regions). The mother guards the nest aggressively for about twenty-eight days, and then the goslings hatch—usually on the same day. The young birds are almost full sized within a year.

There are several problems with these big birds. Canada geese are grazers, so they are *very* social birds. When they move into an area, it's as if all of your relatives and distant relatives decided to camp out in your backyard at the same time. Unfortunately, Canada geese seem to prefer grasses and lush green vegetation, particularly in urban areas. Now many of our cities' parks and other green spaces (including golf courses!) are filled to capacity with these honkers—particularly if they are in waterfront areas. Meanwhile, local farmers have to put up with the geese feeding in their cornfields.

Studies have shown that Canada geese have adapted extremely well in our cities, partly because food is plentiful, but also because, in an urban setting, there are few natural predators of these birds. City planners are further alarmed when they learn that these geese normally live about twenty years. And since geese tend to use the same nest year after year, it's hard to evict them once they've moved in.

This invasion creates other serious problems. Geese are voracious eaters; they will strip vegetation bare. At the same time, they generate enormous amounts of green slimy waste. This goose dung then kills grass, creates an awful odor, fouls parkland, and poses serious health problems. When the land is adjacent to a pond, lake, river, or stream, the goose droppings leach into the soil and frequently pollute the water supply. In Toronto, for example, sections of Lake Ontario beaches are routinely closed each summer as a result of goose bacteria problems.

If the geese are already established, it's possible to control them by habitat modification. Some cities restrict geese with extensive fencing, or prohibit the public from giving them handouts in certain areas. If natural barriers are added—such as hedges, shrubs, prickly bushes, or stones or fences—geese can be

Canada goose (*Branta canadensis*)

blocked at the water's edge. When confined, they may feel more threatened than they would in open areas where they can easily detect any predators.

Homeowners can discourage geese from grazing on the lawns by switching from the popular Kentucky bluegrass to a taller and less tasty variety. Other ground covers that irritate them include hosta, various kinds of ivy, and myrtle. And if none of these techniques are helpful, here are some other preventative measures.

Knox, a pest control firm in Chicago, uses border collies and mute swans to keep geese away. The dogs harass the geese into leaving, and the swans aggressively defend their territory and also drive away geese. Sound techniques also work in some areas. These include whistles, screaming machines, crackers (small explosives), and various banging devices. Visual devices are successful, too. Some parks use kites that resemble eagles flying overhead.

If these tricks don't help, the MiniScare Windmill is a unique device that deflects ultraviolet light off its blades to repel the birds. If you still find geese chomping in your front yard, try some of the commercial fertilizers that help your grass grow but apparently taste disgusting to Canada geese. Some of these concoctions, including Bobbex-G, actually have an ingredient that makes the repulsive taste stick to the goose's beak. One bite and they're not likely to dine on that lawn again.

BATS

Bad reputations notwithstanding, bats are actually pretty good critters. They eat mosquitoes, moths, small rodents, and other unpleasant little pests. A typical female bat will kill six hundred mosquitoes and other insects in one evening. It's estimated that the bats around San Antonio, Texas, eat one million pounds of insects every night. Now that's some insecticide! Nonetheless, bats still have a public relations problem. We've harbored many mistaken notions about them for a long time. But here are the facts:

- Bats are rarely rabid. Estimates say that only .5 percent of all bats carry rabies. You have a greater chance of contracting rabies from a cow.
- Bats are not blind. Not only do they actually see quite well, but many bats also use a sophisticated sonar system that helps them locate their prey and zap it at high speed.
- Bats do not fly into people's hair. They are more scared of us than we are of them, and their sonar prevents them from making aerial mistakes such as heading into a head of hair.
- Bats are not disease spreaders and humans rarely contract the few diseases that they do carry. There's only a small possibility of picking up histoplasmosis, which grow in bat guano or in soil enriched with bat or bird excrement.
- Bats do not bite people (unless they are being handled), and they do not suck blood. Although there is a variety called vampire bats, they only suck the blood of livestock, and no vampire bat can be found in North America.
- Bats live just about anywhere, excluding the polar regions. They have been tracked at altitudes as high as ten thousand feet, about the elevation of the White Cloud Mountains in Idaho. They generally fly in the slow lane, although some bats can travel at thirty or even sixty miles an hour.
- Bats are the only mammals that really fly; flying squirrels are actually gliders. The Greek name for the bat order is *chiroptera*, which means "winged hands." The bat's arm and finger bones are greatly elongated to support the elastic skin and muscles they use to fly.

TOTAL CRITTER CONTROL

Bats basically evolved along two paths: the fruit- and nectar-eating variety (sub-order Megachiroptera), and the insect-eating variety (sub-order Microchiroptera), although 85 percent are in the latter category. Nectar- and fruit-eating bats are farmers of a sort. They drop seed-laden guano throughout the Americas, and nectar-eating bats help pollinate tropical flowers. Without bats, it is possible that many species of plants would simply disappear. In fact, if it weren't for bats, you could forget about ever enjoying a margarita or a tequila sunrise. Bats actually pollinate or disperse seeds for more than three hundred plant species in the Old World tropics. You won't find any nectar- or fruit-eating bats in North America, but we have forty-two species of our own bats. The most common is the little brown bat (*Myotis lucifugus*), and it's actually a cute little devil.

Insect-eating bats survive on a diet of bugs we love to hate: mosquitoes, caddis flies, moths, and beetles. To home in on these tiny morsels—and to navigate at dusk, dawn, and occasionally at night, when insects are active—bats have evolved an adjunct to their sight. Their ace in the hole is the ability to echolate, that is, to use sound to "see"—even in complete darkness.

They bounce high-pitched squeaks (inaudible to most of us most of the time) off an unsuspecting insect, capture the return echoes with their oversized ears, and then perform intricate calculations to arrive precisely at the same spot at the same moment as their prey. They actually capture the bugs in their wings and then transport them to their mouths. Any bat sounds we may be able to hear are the audible squeakings, chatterings, and clicks that they use to communicate among themselves.

Most bats we encounter are social and live in colonies that sometimes number in the thousands. The Mexican free-tailed bat (*Tadarida brasiliensis*), found in Mexico and in caves in the American southwest, may form colonies of more than a million individuals. Most bats mate in the late summer and fall but don't give birth until the spring. During the summer, the female bats and their offspring gather in large maternity colonies. (The males usually live solitary lives while the young are nursing, and some species form bachelor colonies. The stray bat you find behind the shutter is probably a male.)

Most mother bats have one offspring per year, which weighs about one-third of the mother's weight (that's like a 130-pound woman having a 43-pound baby). Mothers and babies keep the nursery temperature cranked on high—temperatures can reach over 130 degrees Fahrenheit—because the heat stimulates growth. After three weeks, the baby bat is ready to fly.

Mother bats are very devoted to their children. The only time mothers leave their babies is to go food shopping at night. What is amazing—at least to us—is that when they return to their cave, mother bats are able to find their own offspring among all the other bats. Once the young bats are weaned in late summer, the nursery disperses.

The Winged Critters

If a human or other intruder disturbs these maternity colonies, the effect can be devastating. Bat moms may be forced to abandon the offspring or may try to transport them, with deadly results. If the nursery is disturbed, the mothers will attempt to move the young by attaching them to their teats. Vandals—the human kind—will often kill the harmless young bats, and the disturbed hibernating bats that are forced into the cold air will use up all of their stored calories trying to keep warm, and die. If the baby remains unmolested, it matures quickly and is soon flying and hunting with the rest. Some of the forty-two species of bats in North America are endangered, so be careful around potential bat habitats.

Insect-eating bats emerge from their roosts in the evening and usually head straight for water. They drink by skimming the surface. Next, they feed on insects for about half an hour, eating until their stomachs are distended with the fare. They rest then while they digest their meals. Mothers will return to the roost to feed their young, but others will find a temporary open roost under a porch, bridge, or tree. They'll feed again before dawn.

Although bats are mobile at night, they're vulnerable during the day—as well as during periods of hibernation—when they roost in caves, trees, or attics. They're prey to many animals: owls, hawks, cats, raccoons, and snakes all eat them. As mentioned earlier, vandalism and repeated disturbance of roosting caves is a primary cause for a drastic decline in bat population numbers. Forty percent of our native bats are on the endangered species list or are official candidates for it.

Despite the hazards, bats live a long time for wild animals, about thirty years—a result of their energy conservation adaptations. Active bats have a body temperature of around 100 degrees Fahrenheit. Roosting bats, unless they're in the nursery, save a little energy by dropping their body temperature to equal that of the environment, thus conserving energy. To survive cold winters, bats either hibernate or migrate to warmer climates.

Like many other animal species, bats are running out of suitable habitats—forests or loosely constructed outbuildings—which increasingly brings them into contact with humans. After all, you like your house, why shouldn't bats? In fact, during World War II, U.S. military experts were so sure that bats would seek out roosts in houses that they drafted the little critters. In project X-ray, Mexican free-tailed bats were equipped with small incendiary time bombs, refrigerated (to calm them down), and then dropped over enemy territory, particularly Japan. Fortunately, the military soon abandoned this outlandish project.

Bats are most likely to bother people in August, when they are just starting to explore the world outside their cave (or tree, or attic). We can appreciate the point of view of people who would rather not have bats around, especially in their attic. Indoor bats are scary and leave guano all over the place. A large accumulation of bat dung can be toxic, occasionally causing severe respiratory problems. People have devised a variety of methods for getting rid of nuisance bats. About three dozen bats were trapped in a Salt Lake City, Utah, office building, for example. A security guard herded the bats down twelve flights of stairs using a broom!

That's one way to get rid of them. Others include loud music, mothballs, and bright lights. Use five pounds of mothballs for about two thousand cubic feet; you can hang them in mesh bags or spread them on the floor and on the walls. Arthur Greenhall of the U.S. Fish and Wildlife Service reports that some people have successfully used dog whistles hooked to oxygen cylinders or large aquarium pumps to repel bats. But ultrasonic devices have had little effect, he says. Bats can get used to almost anything, so these solutions are iffy. Still, plenty of bats live in our homes without ever causing problems or even being detected.

In most areas, bats are considered a non-game species so you can't hunt them, and poison is also outlawed in many regions. Besides, even if you kill them, you're still providing a comfy bat home, so new bats may eventually replace the old, dead ones. As mentioned above, some bats are also protected by endangered species laws and, at any rate, officials usually frown on the killing of any bat.

As a matter of fact, the University of Arizona in Tucson faced charges related to killing bats. When they decided to get rid of a colony of bats roosting in the concrete football stadium, university personnel sprayed the roosting mammals with carbon dioxide from a fire extinguisher, both freezing and suffocating the animals. Then a local pest control operator followed up by applying a sticky bird repellent containing polybutene to the roosting surfaces, and the remaining bats had their wings glued to their bodies. The pest control company was fined by the state for improper use of a pesticide.

If you find a bat in your house, don't worry. Bats are singularly uninterested in biting or even landing on you. Just be sure to stay calm, open some windows and doors, and the critter will eventually find its way out. The best way to keep bats out of your house in the first place is to eliminate any openings through which they might enter. Unlike other mammals, such as squirrels and raccoons, bats won't chew their way into a house. But they will squeeze through the tiniest of holes, so you need to be diligent in finding and eliminating such entry points.

Patience plays a role in getting rid of bats. As night begins to fall, stake out a comfortable spot to observe the bats as they exit your building, then block the holes. Becca Schad, owner of Wildlife Matters, a Virginia pest management firm, says, "A lot of times in the summer I work at night because that's when the bats are active. I go out to a house and start looking around about fifteen minutes or so before it starts getting dark and the bats become active and fly out. By noting where their exits are, I can go behind them and close up the exits."

Sounds easy, right? Blocking up the easiest bat access, however, doesn't mean they won't find another hole. So before you block the active access, close off potential ones. One way to find potential bat doors is to seek out places where the air flows from the house. You can find air leaks with inexpensive household items: incense, tissue paper and a hanger, or a candle. Tape the tissue paper along the bottom stem of a clothes hanger and hold the device near eaves, walls, and windows you suspect are leaking air. Smoke flow from incense or a flickering flame also indicates air leaks.

Some spots are obvious. Schad says that bats often enter through the louvered vents in an attic and she recommends using one-quarter-inch hardware cloth to block the access. Vents, spaces around doors and windows, loose screens, chimneys, and cracked roof flashing are other likely spaces.

The best time to bat-proof is at the end of the summer when young bats are ready to fly and they're not yet ready to hibernate. After you close up the auxiliary holes, block the main exit. After you block up the main exit, open it on several successive nights to make sure all of the bats got out. Repeat this process as necessary.

If it's awkward to close off the final exit after dark, there are a number of one-way doors that can be used temporarily until the bats leave for good. Some of them collapse after the bats fly out and others use a cone design where bats exit from the small end of the cone and can't fly back in. Perhaps one of the easiest, temporary, one-way doors are bird netting strung across the main access. The bats leaving the building climb down the netting, drop, and fly. On returning, they would usually fly right into the hole, but the netting blocks their path. You can make permanent repairs with caulking, weather-stripping, metal flashing, screens, or insulation. You can also block holes under roofing with fiberglass batting or rustproof scouring pads.

Schad adds, "More and more these days people are realizing that bats don't deserve their horrible reputation and that they really are beneficial because they eat insects." Even so, many people are still pretty squeamish about actually going up in an attic where they know there are bats, so "a lot of times, they call me," says Schad. Yet, she says, bats "really are one of the more harmless creatures that get into houses."

Sometimes people discover bats hibernating in their attic during the winter. Don't disturb the bats then, because you can't remove hibernating bats without endangering them. If you oust a bat in winter, it may well die of starvation because its insect prey is no longer present. Also, bats awakened from their winter hibernation will burn off their stored fat reserve trying to keep warm, and they may not be able to survive until spring.

Another reason not to remove bats in winter is that even if you did wake a few up and get them out, you couldn't be sure that you had gotten them all; bats are good at hiding in cracks and crevices. Also, avoid sealing off your attic when bats are hibernating inside, because when the bats become active in the springtime, they'll be trapped inside and die (or they may try to make their way down into the rest of the house).

CONTROLLING BATS

- Cover louvered vents in the attic with quarter-inch hardware cloth.
- Block up any holes in your house, especially near the roof and eaves.
- Never disturb hibernating bats.
- Make sure you don't trap bats in the house when you attempt to exclude them; they're great at hide-and-seek.
- If all your efforts to remove them fail, call an expert.

Hibernating bats won't hurt anything. Guano won't accumulate, and the pest insects that come along with bats (bat fleas, ticks, mites, and bedbugs) rarely bite humans. So hold out until springtime, and don't disturb the animals when you go up into the attic for your skis. One fine, spring evening you can watch the innocent little critters as they flutter into the sunset. And then you'll be able to scurry upstairs to prevent them from returning.

If you need to clean up after some bat visitors, first dampen the guano with water before removing it. Wear gloves and a face mask or dust respirator while you're doing the work. Spores in the guano (or in chicken or bird droppings) may cause histoplasmosis, a disease that, while rarely serious, usually causes headaches, fever, and sometimes a cough and chest pain. The disease is cured with antifungal medication.

A bat in your living space is a different problem. If it's night, turn out all the lights, and try to isolate it in one room by closing off the rest of the house. Failing that, when the bat lands, try to cover it with a coffee can or cardboard tube (it will probably be grateful for the cover). Slip a piece of cardboard under the container to seal it, and then take the container—and bat—outside to release it. You can also catch a bat with a butterfly net, in a towel, or with leather-gloved hands, but remember that even the gentle bat may bite in self-defense.

Just try not to panic. Remember that the bat doesn't really want to be in your living room. And it will be happy to leave once it spots an opening.

And the most important tip of all? Give bats alternatives to settling into your home when you oust them. Bat houses, which look like oversized bluebird houses, are available in many nature and wild-bird supply stores, and through a number of catalogs.

SCALY ANIMALS AND WATER CREATURES

ALLIGATORS

The North has cold, snowy, icy winters, which many people regard as a draw-back. But the South, the Gulf states in particular, have alligators. On a National Geographic special, they're fascinating; without the television screen for protection, they're frightening. When they aren't behind a protective wall in zoos and adventure parks, they can be downright scary.

Here's just one example. A boy was leisurely riding his moped along a road in Travis County, Texas, when a four-and-a-half-foot alligator blocked his path. Not your everyday roadblock, but a mere baby as far as these creatures go. The police were summoned, and the two deputies who arrived hog-tied the alligator, which managed to escape the ropes twice before the deputies got the knots in place on the third try. The alligator was transported to a nearby lake, where it was set loose.

Or maybe you'd rather take a swim in Long Beach, North Carolina? Alligators prefer fresh water, but when ponds and lakes become shallow, gators seek out water wherever they can find it. One August in recent years, an alligator made its way to a heavily touristed section of the ocean at Long Beach and spent several days playing hide-and-seek with tourists and police. An officer involved in the hunt said the police were only able to capture the gator after they "slipped a four-wheeler between him and the water and cut him off."

Then there's the story of Adolph in Indianapolis. It seems this four-foot, nine-inch gator escaped from the Old Indiana Amusement Park and stayed away for seven weeks. Teenagers who thought it would be fun to watch this critter crawl around had freed Adolph. He eventually showed up at a park lake, where an animal trainer spent a month trying to lure Adolph back with chickens. No luck. They finally had to drain the lake to trap him.

As if it's not bad enough that the outdoors is crawling with these things, some people have acquired pet alligators. Evidently, they make great watchdogs; *their* bite is certainly far worse than their bark. But if you ever try to get rid of an alligator, *that's* not so easy. One Colorado resident who advertised his three-and-a-half-foot "Wally" in the newspaper received only one inquiry. When the potential buyer found out that Wally would eventually reach six or seven feet, he slithered out of the deal.

Native to Southeast wetlands, the American alligator (*Alligator mississippiensis*) is our largest reptile and may grow to lengths of eighteen feet. Most of them probably don't reach that size, and today a large one is around twelve feet long. Yearlings may be eighteen inches long, and they grow about a foot annually for several years, under good conditions. So only a dedicated owner would keep a pet gator for very long.

Heavy dorsal scales protect the alligator's back, and the rest of the body is covered with dry scales. Unlike mammals, alligators are cold-blooded, which means they don't have any internal control to maintain a constant body temperature. They prefer warm weather, hot, even. But not too hot. Alligators cannot survive body temperatures above ninety-five degrees Fahrenheit. Poachers who want a perfect carcass let the sun do their dirty work and they simply leave an

American alligator (*Alligator mississippiensis*)

immobilized animal in the sun for several hours until it dies. Left on their own, alligators stay on the move in their territories, always trying to find somewhere warmer or somewhere cooler.

It is the alligator's unending search for something better that often brings them to our yards, resorts, and pools—and into conflict with us. The bad news is that they like many of the same habitats that humans do: sunny waterfronts, golf courses, ponds, and canals. Any fresh water, and even some brackish or salt water, is suitable for gators. The good news is that they're territorial, so usually you only have to deal with one at a time.

Alligators are unrepentant carnivores. They have no worries about heart disease or cholesterol. They eat lots of meat, and they're well equipped for their diet. They have sharp, conical teeth and several replacement sets of them at that. But the teeth don't regenerate indefinitely, and an old alligator may be nearly toothless. (Unfortunately, it may not be possible to determine the age and condition of the gator's teeth until it's too late.) Under the very best of conditions, an alligator may live to be fifty years old.

Alligators eat fish, amphibians, some birds, other reptiles (especially snakes and turtles), carrion, and small mammals. Little particles of blood or flesh in the water often give an alligator the first indication of nearby food. When they sense these appetite stimulants, they'll violently lash their heads from side to side, snapping their jaws, looking for the tasty tidbit. Overlooking the question of mating for a moment, gators place most things in two categories: food and non-food. Generally gators regard people as non-food unless unwise folks cement a human association by feeding them.

Gators are always looking out for their next meal. They're opportunistic feeders and eat whenever they're given a chance. If they are too full to eat something, they'll stash a meal under a log at the bottom of a lake. Just a snack for later. Cold weather makes them sluggish, and in winter they become inactive and feed less often. In some northern parts of their range, alligators go into a sort of hibernation and burrow into the muddy ground to wait out the cold.

With the advent of spring, an alligator's thoughts turn briefly from food. Males and females come together for about three days, for courtship and mating. After they copulate, the male leaves for his own home range, and the female remains in hers to build a nest and lay eggs.

There is little outward differentiation between the sexes, and unless you can get close enough to manually examine the internal genitals (which is not advised) it's tough to know which is which. But a dead giveaway is an alligator guarding a nest; you can bet that's a female. About two months after mating, she builds a nest of mud and whatever vegetation is most handy, about three feet high and five feet wide, and she scoops out a cavity where she will deposit thirty to fifty eggs. The nest is usually constructed in a shady area not far from the water.

Some of the most dangerous alligators are females protecting their nests from intruders; females have reportedly attacked bears that ventured too close to the nest. On one occasion, a female attacked an airboat of Florida Game Commission census takers. Even an alligator has trouble harming a boat, but the officers stayed away, nonetheless. Females will attack almost anything that ventures too close to a nest. A nesting female embodies maternal protection as an armored, snapping, fight-to-the-death fury.

Alligator eggs are about three inches long and the same diameter at both ends. The female does not need to incubate the eggs; in most of the alligator's range, the air temperature is warm enough to do the job. The mother stays in the vicinity of her nest while the eggs incubate. During the day, she often rests on the nest with her throat on top of it. In addition to keeping the nest safe from predators, such as raccoons who would eat the eggs, she dampens the structure with water that drips from her body after a swim and keeps the eggs from overheating.

Baby alligators dig out of the eggs and then the nest by themselves between July and September; the timing depends on weather conditions. Her maternal instincts are strong, and the mother stays around the young for a few months to protect the small reptiles from predators, including other alligators. Even so, it is estimated that as many as 90 percent of baby alligators die within nine months. Juveniles who survive their first year in the wild will probably survive into adulthood.

In natural history parlance, there are limiting factors that keep animal populations in check. One limiting factor for the alligator is habitat. Once on the endangered species list, the alligator is rebounding. Hunting is now regulated and even the environment is a little more benign. Many naturalists think that development of golf courses and canals in Florida has increased the livable habitat for alligators in recent years and allowed the population to expand. But a burgeoning human population has also brought about more interspecies contact than either would prefer.

Florida is pretty much the alligator capital of North America. The Florida Game and Freshwater Fish Commission receives nearly ten thousand calls a year complaining about the presence, or behavior, of an alligator. And about a dozen full-fledged alligator attacks are reported in the state each year. Lieutenant Tom Quinn of the state game commission says that many of these encounters could be avoided. He gives one example: "Three men were driving along when they spotted an alligator by the road. They decided to put it into the trunk, and one man was bitten." In that case, it was hard to blame the alligator.

Bill Brownlee of the Texas Parks and Wildlife Commission concurs. "People's curiosity gets them in trouble. A gator may look dead, when it's not." So don't get

Scaly Animals and Water Creatures

closer to find out. He recommends that people stay at least thirty feet away from an alligator. Texas gets about a hundred alligator complaints each year, and the commission destroys about fifty gators annually.

Given alligators' large size, sharp teeth, and small brains, it seems that we should be the ones to exercise caution, rather than the other way around. Your best strategy for reducing the risk of a gator encounter will vary depending on where you and the alligators are. If you're outside a major urban area in gator country, any body of water could harbor alligators, so assume it does. An alligator that's at least four feet long is a threat, and if you can see one or know one lives in the area, then it's not wise to swim there or to send your dog into the water chasing after sticks.

People with water dogs need to be especially cautious. Mark Westall, a naturalist on Sanibel Island in Florida, explains that when people throw a tennis ball out into the lake, the disturbance can act like a feeding stimulus, alerting an alligator that a potential meal is at hand. "The alligator looks at that dog and says, 'Well, that's not quite a raccoon, but it looks a little like a raccoon.' " A Texas wildlife official says dogs are like an ice cream treat for alligators. They can't always tell the difference between a retriever's legs and a raccoon's or a deer's—and, generally, they don't waste time trying.

Of course, this naturally brings up the question of whether they can tell the difference between a deer's legs and your child's. Accounts of alligators grabbing dogs naturally make some parents nervous. "They think, 'Well, if the gator is going to take a dog, which weighs ninety pounds, what about my children?'" says Westall. He maintains that alligators look at dogs and say, "There's a funny-looking raccoon." But alligators look at human beings and say, "Here is a predator, here's a killer who is going to kill me." Westall says an alligator naturally wants to keep a respectful distance.

But all too often the alligator has learned that people mean food—garbage, fish entrails, or even purposely fed snacks. An alligator unfamiliar with people will be ignorant of our food possibilities—but what are you going to do, ask it? "Excuse me, Smiley, but has any human ever happened to feed you?"

Natural bodies of water may not be the only ones harboring alligators. When it gets really dry, alligators may cast a longing eye at our swimming pools and man-made ponds. When natural water is sparse, alligators head for any swimmable source of water. Even ponds as deep as twenty feet are attractive to alligators. From a gator's perspective, water is water. For an alligator, a swimming pool is a perfect place to run and hide when confronted by a human. Alligators do not jump into pools to eat people; they go into pools to escape people. It matters little whether there are swimmers in the pool already; fresh water is a safe place for an alligator. Land is not.

TOTAL CRITTER CONTROL

Although alligators will venture into pools for protection, they don't like the chlorine. So, left on their own, alligators will get out of a swimming pool as soon as they feel it is safe to do so. Of course, it's easier if there are steps in the pool or some contrivance placed there that helps the animal to exit—kind of like putting a board in a basement window to allow a small mammal to get out. Clear the area and the alligator will leave. Most of the time.

Mark Westall tells a harrowing tale of trying to get one out of a subdivision's pool. "He sank to the bottom of the pool, so I sank a rope and slipped it around his neck. I pulled him out of the pool and started dragging him across the fairway to let him go in the lake." But things didn't go according to plan.

"An alligator is like a mule. The more you pull them in one direction, the more they want to go in the opposite direction. They think you're dragging them to their death. Why else would a human being put a rope around their neck? So, I'm dragging him with about fifteen feet of rope, and about halfway across the fairway, the rope went limp. Before I could turn around to see what had happened, the alligator was brushing my legs as he ran by; he was about an eight-footer. If I was bowlegged, he would have run right between my legs.

"And the next thing I know, the alligator is leading me to the water. The alligator saw the lake and saw safety. He could have easily grabbed me if he wanted, but he wanted to get to the water. I looked like I had a Doberman on the end of the leash. I loosened the noose, and he did a racing dive into the lake, and that was that."

The water isn't the only place you might encounter a gator. Alligators also like to move around a bit, so you may meet one on its way from one place to another. Where are the alligators going? "They used to live in a marsh, now it's a golf course; they're going to move into the pond there. Also the other thing is alligators naturally are going to move from pond to pond—especially in the spring when they are trying to mate," explains Westall. And each yearling needs to find a home, too. They will also be on the move after the heavy spring rains or during floods.

With today's subdivided world, ponds and blocks of land are geometrically and pleasingly arranged for developers and architects, not for the ease of alligators. In the old days, alligators would go from a low area, walk through a marsh, and get to another low area; today they have to get up and walk across areas that we've filled in to make land for houses. When alligators want to leave their pond to try to find a mate in the pond across the street, they walk across property. "If an alligator is sitting in your driveway, he's not there to hunt your kid, he's just trying to get to the pond across the street," Westall explains.

Westall is especially tolerant of gators and, with some other naturalists, actually initiated a program to provide "escorts" for Sanibel's roaming reptiles. "It's the only place in the entire state that has such a program," he says. The escorts were once a group of naturalists, which included Westall and some others who were licensed by the state to move alligators around. Today the responsibility is handled by the local police force, trained by Westall and the other gator guardians, although Westall still escorts or moves an occasional alligator. The escorts are there to keep the gator out of trouble, not guide it to any one location.

The wary tolerance of Sanibel Island helps to keep conflicts at a minimum there. As Westall explains, "On Sanibel we realize the alligators are going to walk around periodically. If you live here for a while, you get used to it." Problems grow, however, as more and more newcomers arrive in the area—people who "freak out when they see an alligator walking down the street." Yet, for the time being at least, escorts seem to work, and if you have regular alligator crossings in your area, consider implementing this kind of escort plan. But first check out your plan with your state's fish and wildlife agency, since contact with alligators is regulated. Local naturalists and outdoors clubs are good places to find help, too.

But what if some Sunday you decide to mow the lawn and there's a big gator parked right in the middle of the yard? You could use it as an excuse to go back inside and watch the ball game, or you could go on the offensive. (Actually, that gator should have scrammed for cover as soon as it heard the screen door slam.) Clear your throat, yell, and then wait a bit. Chances are if it didn't scoot when the door slammed, it'll leave at the yell. Most alligators have a healthy fear of humans; we nearly wiped out the entire species to make a fashion statement, and they haven't forgotten it.

An alligator on your turf that *doesn't* flee has lost its fear of humans. Removing it is a job for professionals, so call your state's wildlife office. Or maybe the animal is a female with a nest, in which case you should see the mound nearby. Is this a nuisance and a danger? Or is this an unparalleled opportunity to observe untamed nature? If you adopt the latter attitude, just bear in mind that you'll lose the use of your yard for about five months. Every year.

Waterfront houses are most likely to have frequent gator visitors. But if you can keep the animals out of your yard to begin with, you can mow the lawn whenever you want. The key is to make it difficult for the alligator to come up from the water into the yard. Heavy vegetation should work.

People who clear all the vegetation along the waterfront, which many people do because they want to plant sod lawns, actually create attractive spots for alligators, a regular Coney Island for the green set. Clearing the lakefront is a little like putting out a bird feeder and then getting upset because birds visit. Leave existing vegetation or plant new vegetation along the lake; plants such as spartina, cord grass, and leather fern will create a thick, grassy, or fernlike buffer along the waterline.

You're no doubt thinking, "Well, the alligator will walk right through those grasses." But the alligator doesn't want to bushwhack through the grasses to get up to the rest of the lawn. It's hard work, and it cuts off a speedy exit. Shrubbery prevents them from swiftly moving back into the water, their refuge from encroaching humans. As naturalist Mark Westall says, "Remember the old *Tarzan* movies, when the crocodiles would dive off the bank? Notice how they weren't running through any major stands of vegetation. They were sliding right off the bank into the water." Plant some vegetation that creates a buffer between the wild area and your place, and the alligators won't sun on your lawn.

Nevertheless, this natural solution has a few drawbacks. While the vegetation strip might keep gators from sunning on your lawn, it also offers something that alligators find attractive—which is cover. If the vegetation is located on the edge of your property, that's where the gator will stay. The cover might provide a good place to hunt or even to lay eggs. But, if you don't want to go this route, there's another option, according to Mark Westall.

"On Sanibel we let people put up a fence, because a lot of people trust man-made structures more than they do natural systems." If a fence is put in, it should be sunk in the ground a bit to keep alligators from going underneath it. Sanibel's fencing regulations also require people to set the fence six feet back from the waterline to give alligators a sunning spot. "We have a lot of people who want to put a fence in right up against the edge of the water because they don't want the alligator to sun there at all," says Westall. "We think the wildlife has a right to be here, also."

In addition to setting fences back from the waterline, Sanibel residents must also set fences in from the side property lines. Residents have to leave a three-foot corridor, an alligator alley, so that the alligators—and the turtles and the moorhens—have a corridor they can follow to other habitats nearby.

If you do have an unfenced yard, "Don't leave your dog tied up alone in a yard when there are alligators around," points out one member of the Sanibel Island police force. It seems alligators have a fondness for dog meat. Dogs also aren't very savvy when it comes to alligators and may just invite trouble. They'll bark, run around them, even try to bite them on the nose. All the while the gator is waiting for its chance to chomp.

Cats seem to be a bit smarter. They generally steer clear of the reptiles because they don't like the water that much, and despite a reputation for curiosity, cats don't have many questions about alligators. They see an alligator as a potential threat. As a matter of fact, the only time Westall has heard of a cat taken by an alligator was when one was tied up in a yard.

You don't have to live with a nasty reptilian neighbor. States with alligator residents have special programs for nuisance alligators, and they'll send a gator expert to your home to decide how to handle the situation. Often they'll try to relocate smaller animals, but they may have to destroy aggressive alligators or larger animals they can't trap.

Scaly Animals and Water Creatures

Alligator encounters can also occur after dark. According to Mark Westall, "Alligators do most of their walking at night; they're nocturnal, and they walk under cover of darkness. Just like any good commando, they want to stay out of everybody's way if they can." Luckily, there are some simple steps you can take to help you both keep out of one another's path. For example, a flashlight is effective for spotting alligators. Like most animals, gators' eyes glow when caught in the light. Adult alligators' eyes glow red; young alligators' eyes glow yellow.

If you take a flashlight, you can spot the alligator at a distance of about fifty feet. Then you can turn around, go back to your house, call the police, and let them escort the gator out of the area. Or, you can walk over to the other side of the street and continue on your way. Alligators don't hunt like lions, running things down on land. Alligators are only trying to cross over the land. (That, of course, doesn't mean that it won't try to defend itself if you step on its back!)

If you don't carry a flashlight, the alligator is going to see or hear you coming first, and it's going to sit there and remain calm. It may let you come within three to six feet before it hisses at you. The gator hisses because it's scared but, of course, you're going to have a heart attack. All the more reason to carry a flashlight.

More likely than not, if you encounter an alligator in your yard or elsewhere, it will either freeze or turn tail to get away from you. Frank Godwin says, "Alligators are pretty shy animals, really." He should know. Godwin is president of the Gatorland Zoo, a longtime Orlando attraction. Only occasionally will an alligator aggressively approach a human; these few gators think we're a source of food—or even the food itself.

Problems come when people try to use food to overcome the gators' natural shyness. A Louisiana country club's experience offers a case in point. The club acquired three alligators for its lake. Every afternoon an audience gathered to watch the golf pro feed the toothy, open mouths. Soon enough the gators were leaping out of the water to gulp down chip shots. What would be next? A caddie? The sad ending to this story is that the animals had to be destroyed. No one wanted them on the course, and it was unsafe to relocate them to the wild once they had begun to associate humans with food.

Tom Quinn says alligators that attack humans have lost their fear of us, and that's usually because they've been fed. And whether they're fed marshmallows by tourists or scraps of fish dropped at a fish cleaning station, such alligators will learn to associate people with food. In a gator's mind, people practically *are* the food.

Time and again, wildlife officers discover that the aggressive alligators they have to destroy were once nine-inch-long yearlings fed by some well-intentioned person. The lesson is: Don't feed alligators of *any* size or age. This goes for *all* wild animals (okay, except birds). It's serious business, says Westall. "So if you see somebody in your neighborhood feeding an alligator, don't say, 'Oh, that guy's just taming an alligator.' Say, 'That guy may be causing some child to die in this neighborhood.' Feeding alligators is that type of threat."

The problem with alligators is not with the alligators, but with people. We generally do not like to have large carnivores around where we live. Instead of adapting or moving, we kill the species. We have laws protecting alligators, but it's certain that as more people move to the Southeast, there will be increased pressure on these reptiles.

Meanwhile, in case you've wondered, there are no alligators living in the subways of New York City. The giant rats ate them.

SNAKES

A lot of people don't like snakes, but most of these scaly critters are no bother to humans at all. Sure, a snake may curl up with you in your sleeping bag, but most adults can easily escape a snake that shows an interest in getting close.

Statistics help to provide some perspective. Of the world's more than 2,000 different kinds of snakes, only 250 species are venomous. In North America there are only 4 poisonous snake species. So, while there are several thousand cases of snakebite every year in North America, there are relatively few fatalities. Of a thousand reported snakebites, only 3 percent are usually fatal.

Snakes are found throughout the world, with the exception of the cold polar regions. Like all reptiles, they are cold-blooded, unable to internally regulate their body temperature. Instead, snakes regulate their body temperatures behaviorally, by traveling to cooler or warmer locations as survival demands. Like an alligator, they'll die of heat stroke if immobilized in the sun for too long. They live in a variety of habitats from desert to rain forest, treetops to underground. In colder climates, they must hibernate during the winter and, occasionally, snakes of the same or even mixed species will hibernate in the same den, which gives rise, among people with overactive imaginations, to horror stories about nests of rattlers. These nests protect hibernating snakes that will disperse when the weather warms up.

Snakes' internal organs have evolved in some interesting ways in order to fit inside their long, slim bodies, although the adaptations are more pronounced in some species than others. For instance, most snakes have only one lung, the right one, or have only a vestigial left lung. Internal organs—the kidneys, liver, and reproductive organs—are elongated and staggered in their placement.

Snakes are covered with scales, some of which are specialized. A transparent scale covers the snake's eyes, in place of eyelids, and these scales fall off each time the snake sheds its skin. Prior to its sloughing off, the skin becomes cloudy, and the reptile has difficulty seeing. They're particularly edgy around this time because they are vulnerable and therefore they are especially likely to bite. You'll recognize a shedding snake because of its generally tattered appearance and cloudy eyes; this is no time to make its acquaintance.

Although they can't hear or see very well (they're actually deaf but can sense vibrations), they do have an acute sense of smell. The snake's forked tongue darts out, collects particles, and then is withdrawn and transfers the particles to ducts in the roof of the mouth called the Jacobson's organ. Their tongues and Jacobson's organs function much like our noses and are used to find prey.

Snakes are hunters; some actively seek prey while others merely lurk in a likely spot and wait for their prey to arrive. Some snakes have special heat sensors that help them find warm-blooded prey, even in the dark. It seems a bit primitive, but they all swallow their prey whole. Snakes can even swallow prey larger than their heads—the bottom jaw is loosely connected to the skull and can be spread apart. Their teeth are inclined backward, down the throat, to help move the prey down the digestive tract. Although these teeth are often broken, snakes can grow new ones, and even new fangs. Depending on the size and species, snakes eat a variety of foods: eggs, insects, frogs, crabs, small mammals, or other snakes. Some specialize, but many are generalized feeders.

Smaller prey may be swallowed whole without first being killed. But larger snakes, which eat larger prey, must kill their prey first so it doesn't create a ruckus going down. They either squeeze the prey to death or use a venom that kills it. Even non-venomous snakes have exceedingly strong saliva that is designed to help break down complex organic structures—such as animals—so even a bite from a non-venomous snake may be painful. One meal may last for several weeks if the snake isn't very active.

Breeding usually occurs once a year, but because snakes are solitary, finding a mate can be tough. There are no singles bars for snakes. Using their sense of smell, males find receptive mates by following a trail of secretions left by the female of the species. Courtship varies among the snakes, from rather tame body rubbing between the male and female to ritualistic combat dances between males fighting over a female.

Snakes either lay eggs (they're oviparous) or carry them within their body until they are ready to hatch and then produce live young (viviparous). Oviparous females lay eggs in clutches and leave them to incubate, although a few species do guard the nest. Their clutches tend to hold between thirty and forty eggs. Snakes that bear live young may have four to fifteen babies. Viviparous snakes may live in colder regions than oviparous ones, but the mothers pay the price of being heavier, slower, more vulnerable to predators, and less able hunters.

TOTAL CRITTER CONTROL

Snakes are no fonder of humans than most humans are of them. The occasional snake around your garden or yard will make a speedy exit as soon as it knows you are there. But just because the snake hits the road doesn't mean it leaves the neighborhood. Most stick around an area for several months and then move along. They're there because they're busily eating animals that could be pests for you, most likely rodents and insects. That's why snakes aren't usually a threat to humans. We're simply not the right size to be a meal.

There are four poisonous snakes in North America: copperheads, rattlesnakes, coral snakes, and water moccasins. While venomous snakes are found in every part of the United States except Alaska, Hawaii, and Maine, it's unlikely that you'll find any of them in your house, with the exception of the copperhead, which is fond of cool, damp places, such as basements. Any snake is attracted to the kind of cover usually found around houses.

The copperhead (*Agkistrodon contortix*) ranges from Massachusetts to northern Florida and west to Illinois and Texas. Although its bite will make you sick, it's unlikely that it will kill you. Many species of rattlesnake are found throughout the rest of the country, from the desert sidewinder (*Crotalus cerastes*) to the timber rattlesnake (*C. horridus*) in forests of the southern and eastern United States.

The many types of rattlesnakes vary in color. Most are various shades of brown, tan, yellow, gray, black, chalky white, dull red, or olive green. Many have diamond, chevron, or blotched markings on their backs and sides. As do all pit vipers, they have an elliptical eye pupil and a deep pit on each side of the head midway between the eye and nostril. Learn to identify the rattlesnakes common in your region of the country; don't bother with the others. Of the various types of rattlers, the most deadly is the big diamondback (*C. adamenteuus*), whose bite kills fifteen out of every one hundred people bitten, according to snake experts.

Eastern coral snakes (*Micrurus fulvius*) prefer the warm climes of the south from South Carolina to Florida, along the Gulf states and Texas and west into New Mexico and Arizona. Another rarer species, the Sonoran or western coral snake (*Micruroides euryxnathus*), is found in the American Southwest and northwestern Mexico. Both species of coral snake are ringed with red, yellow, and black, with the red and yellow rings touching. Other snakes mimic the coral snake's appearance with red and yellow rings, with black rings separating them. Remember: "Red on yellow, kill a fellow; red on black, friend of Jack." New World coral snakes are relatives of the Old World cobras.

The water moccasin (*Ancistrodon piscirorus*), also known as the cottonmouth, lives in ponds, streams, lakes, and swamps from Virginia to Florida, along the Gulf states and into Texas. Like other poisonous snakes, it rarely bites humans. But for safety's sake, you should assume that any swimming snake is poisonous. Become familiar with local poisonous snakes so you'll recognize one if you see it.

Timber rattlesnake (*Crotalus horridus*)

Local naturalists, park rangers, wildlife officials, or the staff at local sporting goods stores can often provide you with information on snakes in your area.

Bill Zeigler, general curator at Miami's MetroZoo, reports that many of the calls he receives about big, dangerous snakes are really about small, harmless ones. There's a big difference between a water moccasin and an indigo snake—a nearly blue-black, docile Florida snake that can reach lengths of six feet. If you think you've spotted a poisonous snake around your home, report it to the local game or wildlife department immediately. And then keep track of the animal's movements so you can show the official where it went. If you can't point it out, it's going to be pretty tough to find it.

Even though few of us die from venomous snakebites, it's good to know about them. Children are most vulnerable to the bites, so it's best to get the victim to a hospital as quickly as possible. A poisonous snakebite causes an almost immediate reaction: swelling, darkening of tissue around the bite, a tingling sensation, and nausea. All snakes have teeth and will leave teeth marks, but only a pit viper will leave fang marks. Wash the wound with soap and water, elevate the wounded area, and keep the victim calm.

A national parks employee in Virginia relates the wrong way to react: "This guy hurt the girl worse than the snake. He was so sure it was a copperhead bite that he crisscrossed the wound with knife incisions and even tried to suck the poison out. He had her all cut up, and you could tell it wasn't even a copperhead bite." Leave that stuff to the TV movies.

If you have snakes around your house and garden, it's probably because you have a good food source for these carnivorous reptiles, or maybe you have a great hiding place: tall grass, accumulated flower pots, rock piles or woodpiles, heavily mulched flower beds, dense brush, or any damp and cool (or dark) spaces. For instance, a large woodpile on the ground attracts field mice. Snakes, of course, are big mouse-meat fans and will haunt your woodpile in anticipation of a tasty meal. To thwart the snakes, you can either trap and remove the mice or you can raise the woodpile at least fifteen inches above the ground. Animals won't burrow under raised woodpiles, and your snake should curtail its visits.

Don't use poison on the mice and snakes. Why not? If you have a snake "problem," it's probably because you have ample food around, such as mice. Over the long run, poison will do more harm than good. Mice reproduce at an alarming rate, whereas snakes only produce offspring once a year. If you poison the mice, you'll also poison one step up the food chain—the snakes—and you'll still have the mice. But the snakes will take longer to bounce back, and the mouse population will test the limits of the local food supply.

Other likely habitats for mice (or other sources of food for snakes) include high grass, bushes, shrubs, rocks, boards, and junk around the yard. These little rodents also come to feast on spilled birdseed, ill-maintained compost, and your pet's food. What's cover for rodents is cover for the snakes. Snakes love junk. Your old refrigerator out in the yard? A snake mansion. It's full of mice and hiding holes, so get rid of it. And by all means cover or enclose your compost heap.

Another snake attractor is a warm spot in the sun. On cool, sunny days, reptiles often sun themselves on rocks and other open spaces, which is why you so often see them smashed on the road. (Or are they just crossing?) Make the sunning spot unattractive; shade it, clean it up, slant it. They'll go elsewhere in search of a place in the sun, somewhere close by for sure, but somewhere you won't have to look at them.

Scaly Animals and Water Creatures

A Laurel, Maryland, woman found a snake sunning itself on her apartment windowsill. She moved from the apartment, but she didn't have to. If you are finding snakes sunning themselves on a windowsill or porch, find a way to discourage the animal. Install a temporary incline on your windowsill, positioning a piece of wood, metal, or plastic with the incline sloping out and downwards, and the snake won't be able to rest there. But don't be surprised if it shows up somewhere else, although you'll never be sure it's the same snake.

If you *are* determined to get rid of a particular snake, you'll have to spend time following it around. Try to close up its dark and secluded hiding spot; the animal will then move on to another secure hiding spot. Maybe five feet away. It's hard to anticipate where a snake will want to live—in some old flowerpots, underground, or in your rock garden. Don't get alarmed, but even if you think you don't have snakes, you probably do. Small, secretive snakes you never see often live in gardens or wooded areas. They live just about anywhere; there are even sea snakes.

In most regions, snakes are considered non-game wildlife and are protected by law, so don't indiscriminately kill them. Instead, construct a snake-proof fence around your house or a portion of your yard. It keeps out all venomous snakes and all but the most skilled non-venomous climbers.

Once you're familiar with different snake species in your area, you may develop the more tolerant attitude exhibited by those in the know. Dave Sileck, beach activity director at Florida's Don Caesar Resort, says, "Black snakes are in

BUILDING A SNAKE-PROOF FENCE

The fence should be made of heavy galvanized hardware cloth, thirty-six inches wide with a quarter-inch mesh. The lower edge should be buried six inches in the ground, and the fence should be slanted outward from the bottom to the top at a 30-degree angle. Place supporting stakes inside the fence and make sure that any gate is tightly fitted. Gates should swing inward because of the outward slope of the fence. Any opening under the fence should be firmly filled—concrete is preferable. Tall vegetation just outside the fence should be kept cut, for snakes might use these plants to help climb over the fence.

[Source: U.S. Fish and Wildlife Service]

the pool all the time. Usually they're just on their way to somewhere else. We just get the guests out of the pool—they enjoy anything different around the pool."

Generally, though, if you find snakes in your yard, simply leave them alone. They are undoubtedly doing you a service by eating other critters that are truly undesirable—such as mice and rats.

Sometimes, however, snakes will find their way into houses. A damp, cool, dark basement full of accumulated treasures is attractive; so is an attic full of mice and bats. You may feel more comfortable calling a professional: someone from the animal damage control staff, a wildlife official, or a private firm. If you're sure the snakes are not venomous, you can try removing them yourself with one of the following strategies.

Pile damp burlap bags in areas where you've seen the snakes, and cover each pile with a dry bag to hold in moisture. Snakes will think it's a great place to hide. Wait a couple of weeks, and then one afternoon when the snakes are resting among the bags, remove the bags with a large shovel. Put them into a large garbage bag and set the reptiles free somewhere far away. Be sure to prevent reentry of any snakes by closing up any holes providing access to the house. Check the corners of doors and windows, and around masonry, pipes, and electrical service entrances. If you use mesh, be sure it's one-eighth-inch or smaller.

Residents of low-lying areas may meet their native terrestrial snakes during the wet season. These snakes are generally under a foot long and most are harmless. They can't climb trees, so they seek dry shelter under the doors and in the cracks of houses. You may find them under the rug or in a dark closet, and you may find more than one. Since you can't just open the door and yell "Scat!" you'll have to capture and release the animals. Just gently sweep them into a wastepaper basket or other container and release them on high ground outside. Don't pick them up, because their bites can be painful. (Remember, even non-venomous snakes bite.) Weather-strip the doors and fill in the holes to prevent them from returning.

James Knight, with the U.S. Department of Agriculture Extension Service in New Mexico, hit upon a way to trap snakes. He attached some six-by-twelve-inch sticky traps to sixteen-by-twenty-four-inch pieces of plywood with a hole drilled in one end and then he anchored the devices along walls in a dwelling. (Like most animals, snakes don't like to cross open spaces, so they'll tend to keep close to the walls.) When a snake would become stuck to the trap, usually after two days or so, he would retrieve the trap by slipping a hook (long-handled!) through the hole drilled into the plywood. If you try this method, be sure not to place the traps near pipes or structures that can give leverage to a snake trying to escape.

After securing the captured snake, Knight pours cooking oil over it and gently pries it from the trap with a stick or pole. (Cooking oil will free animals from sticky traps and glues.) Knight had observed snakes in his lab for two months to make sure the method didn't harm them. (Snakes shed any sticky residue with their old skins and they suffered no ill effects.) No need to mention that if you're afraid of snakes, this is not the method for you. But it's an effective, non-lethal method provided you check the traps regularly and remove them entirely after a few weeks.

Finally, we should have a quick look at a few "illegal aliens." Most snakes that you encounter in the wild are small and not dangerous, but in southern Florida, that's not always the case. Owners of large, exotic snakes are letting their pets go free in the wild. And a few of these pets, longing for the freedom of the wild, simply escape their owners. Boa constrictors and pythons are favorite species for snake owners, and consequently they are what you may encounter during a picnic. In June 1990, three exotic snakes were captured in Florida: a ten-, a twelve-, and a fifteen-footer.

Nobody knows for sure, but hundreds of these snakes may be loose. Twelve- to fifteen-foot pythons have no difficulty devouring a small dog. That's often the reason why snake owners release their pets. In the beginning, these snakes are cute (well, they are to their owners). Later, feeding these large snakes can become exorbitantly expensive. You run out of poodles, or whatever, quickly.

Unfortunately, exotics such as boas and pythons "are probably going to become a staple around here," according to Bill Zeigler. The exotic pet trade is big business with few restrictions; scores of animals come into our country but most

SUBVERTING SNAKES

- Get rid of junk or move it away from your high-traffic areas. All sorts of animals will live in old appliances, cars, boxes, and other large castoffs. If you can't cart them away, at least move them away from the areas where people walk, play, and congregate.
- Raise the woodpile. Keeping wood stored close to the ground invites all kinds of critters to take up residence.
- Get a cat to help you with any mouse problem. In turn, that should take care of any related snake problem.
- Block access to your house, crawl spaces, and basements. If the door is open, the guests will arrive.

of them aren't tracked or licensed. And many of the tropical reptiles thrive in south Florida's climate.

Zeigler says you'll probably find boas and pythons, both non-venomous constrictors, around urban areas. They're there because we're there, and because we always attract plenty of rodents. While boas don't grow much longer than nine feet, pythons can reach twenty feet. At that size, they're a danger to house pets and even small children. Pythons generally won't eat pets, but they will squeeze them until they suffocate. So don't leave your kid napping alone in the yard because "you'll be stretching your luck," says Zeigler.

In general, it's a bad idea to let your pets roam. Not only does Fluffy make a litter box of your neighbor's garden, but she could also become prey to any number of wandering predators. With the increasing frequency of python sightings in south Florida, there's even more trouble for unattended pets. But pythons or even poisonous snakes are less of a threat than other carnivores such as coyotes.

SEA LIONS

In most California zoos, sea lions are a popular attraction. Their acrobatic antics and doglike barks amuse and amaze children and adults alike. Nonetheless, in some cities where the local zoo proudly displays these creatures, there are city officials on the other side of town who are desperately trying to think of ways to rid their docks of sea lions. Why would anybody want to rid themselves of sea lions, which many people are willing to pay money to see? Well, in San Francisco and Monterey, for example, a proliferation of sea lions has, at times, created a nuisance. Looking for a place to rest, sea lions have decided to relax on boats; when a half-dozen eight-hundred-pound sea lions climb on a pleasure boat, it sinks. Same thing for the docks themselves. Then there's all that sea lion fecal material, which is not insignificant, given the sea lion's size.

You can't kill sea lions because they are protected by the 1972 Marine Mammal Protection Act. You can't move them either. When a tamed sea lion was trapped off the Shilsole Marina in Seattle, Washington, nothing could coax him into his cage. Shove as they might, neither the scientists nor animal handlers could push Sandy, this 615-pound animal, into his cage. He wouldn't move. Hours later, Sandy simply decided that the cage wasn't so bad after all.

The California sea lion (*Zalophus californianus*), like other sea lions, is an intelligent and playful animal. It's most often seen in zoos and circuses, and it can learn an array of tricks, although it's not any smarter than its kin. Sea lions are dark brown and have thin, short, coarse hair. Their front limbs are flippers, and their hind limbs are fused into one large flipper that they can turn beneath them

Sea lion (*Zalophus californianus*)

COPING WITH SEA LIONS

- Moor your boat away from sea lion congregation areas. They're not hard to spot.
- Alter any sunny, flat areas where sea lions gather; make them uncomfortable. Create shade, slippery slopes, and pointy, sharp surfaces.
- Noisemakers such as gas exploders may startle the animals and encourage them to move on.

---

to locomote on land. Male California sea lions grow quite large: six feet long and up to six hundred pounds. Females are about the same length but are slimmer and weigh much less—about two hundred pounds. Sea lions feed exclusively on fish and squid. They breed once a year, gathering on shore in harems of up to fifteen females dominated by one male.

Unfortunately, sea lions can certainly become pests. But if you can't move them, shoot them, or scare them away, what can you do? The only real solution is to wait for sea lions to go away on their own.

ZEBRA MUSSELS

Another aquatic creature that is a real nuisance is the zebra mussel (*Driessen polymorpha*), although it lives in fresh water, not salt water. This tiny critter doesn't have much of a brain, so it shouldn't be tough to outsmart. Yet zebra mussels have managed to cross the Atlantic Ocean and spread throughout the Great Lakes Basin. The mussels originally came from the Black, Caspian, and Aral Seas, and then spread through Europe via the canals in use during the 1700s and 1800s. Sometime in 1986, zebra mussels were sucked up into a ship with freshwater ballast, transported across the Atlantic, and released into Lake St. Clair near Detroit. Then they started hitchhiking around the Great Lakes.

The average female will produce thirty to forty thousand eggs each breeding season, and these eggs hatch into free-swimming larva, or veliger. Veligers can swim about for nearly three weeks before they settle down to the business of attaching to a suitable location. There's even documented evidence that they'll at-

Scaly Animals and Water Creatures

tach themselves to other native mussels and crayfish, and they can settle quickly enough that they'll smother the other animal. They usually like to be in flowing water and often settle where they find currents: water intake pipes, canals, and waterlines. Their colonies decrease water flow, and dead and decaying mussels give the water a horrid flavor.

Susan Grace Moore, a librarian with the Zebra Mussel Information Clearing-house, explains how quickly they can infest waters. "Someone in the Great Lakes complained that a buoy was missing. They expected that it had just broken loose and drifted off. They sent a diver down to where the anchor would be, and they found out that there were so many zebra mussels on the buoy that it had sunk ten feet below the surface."

Mussels can cause severe problems for boats. The veliger settle on boat hulls, increasing drag and decreasing fuel efficiency. They can also be sucked into the water intake of the engine cooling system of the boat. If you turn the boat off and leave it for a few weeks, and you've got veligers in your cooling system, they can settle out and start to grow and create problems—blocking water and causing the engine to overheat.

You can scrape the mussels from shores, docks, and rocks, but be sure to place the debris in the garbage. Boaters pulling out of an infested lake should drain all water, bait buckets, live wells, bilges, ballast tanks, and even the niches of the trailer, and let them dry. Adult zebra mussels have been known to survive in cooled down, out-of-the-water places for as many as two weeks.

Hot water (150 degrees Fahrenheit) can kill veligers and adults, and salt solutions may also be effective. Anti-fouling coatings have been used on boat hulls, but they are relatively expensive and must be reapplied often. Certain compounds are banned in the Great Lakes and may affect other species, so check with your local extension office or the Coast Guard before you use one.

The two-inch mussels don't seem to have any significant biological control in the United States, and no one appears to have found any use for them. It doesn't help that the zebra mussel resembles a native saltwater mussel, and only an expert with a microscope can tell the difference. (That said, it's highly unlikely that you'll find zebra mussels in salt water, according to Moore.) In the meantime, they are a plague on the Great Lakes region, and one of the critters that is proliferating at an alarming rate. You could try eating them, but they're not recommended. "They're quite small, so the energy invested in eating them isn't worth the meat return," says Moore. "Someone who has steamed them has said they smell like a cross between a dead body and fetid gym socks."

9

THE NEIGHBOR'S PETS

DOGS AND CATS

Dogs and cats are wonderful pets—as long as they belong to *you*, of course. But if everyone in the world were a pet owner, then your entire neighborhood would have pets in heat all year round. At that point, the noise and the nuisance might drive you crazy. It would lead to the kind of unfriendly behavior that this famous little story depicts. A woman called her neighbor at 4 A.M. to complain about his dog barking. The man thanked her and hung up the phone. The next morning, at 4 A.M., he called her back. "Madam, I do not *have* a dog," he said.

Disputes between neighbors over pets and pet behavior erupt frequently. Often they end up in court, especially if somebody is bitten or knocked down by a dog. But even the resulting court cases don't always go smoothly. In San Antonio, Texas, an irate homeowner complained that the barking of the neighbor's two dogs caused "psychological pain." The complainant had sprayed Halt (a pepper-based chemical dog repellent) at the dogs to get them to stop making such a racket. In court, the plaintiff's attorney questioned his client about spraying Halt. "Like this?" the lawyer asked, and he tapped the nozzle. Without warning, the spray spattered all over the jury and mayhem ensued. The jurors scattered throughout the courtroom and the judge was forced to temporarily halt the trial.

Animals are part flesh, but they seem to be *mostly* vocal chords. In fact, animal behaviorists have clocked a cocker spaniel that barked 907 times in a ten-minute period. If that barking happened at night, it might have driven sleepless neighbors to do desperate things. In Irvine, California, for example, a neighbor followed through with his threat and killed a dog that kept him awake by barking. Plenty of homeowners recognize the problems and the frustration that dogs can cause, and this is the reason that many communities have laws against dogs that whine or bark for long periods of time.

A control strategy that works in some instances is an anti-bark collar. This little device delivers a shock to the dog when it barks. Most reasonably intelligent animals will quickly get the message, and they'll quiet down. Another possible solution is to install sound-activated sprinkler systems that soak noisy dogs. Most dogs don't like to be sprayed, but these water devices aren't a permanent fix. Dog trainers pretty much accept that it is nearly impossible to train a grown dog never to bark. The trick, therefore, is to remove the stimulus.

Mediation can sometimes help to solve the problem of barking dogs. And we're talking about meditation with the dog's *owner*, of course, not the dog. A Toronto resident, irritated by barking dogs in the middle of the night, turned first to the police. When that didn't help, he went to a professional mediator who managed to work out a deal with the dog's owner. Simply moving the dogs away from the distractions that stimulated them to bark made for a more peaceful night's sleep.

The first step in combating these noisy critters is to keep a record of the barking dog problem. Regardless of the authority to which you appeal, you need to produce "evidence." That usually means you have to come up with proof of the problem's severity. First, present your evidence to the animal's owner. Threaten to go to court if the owner won't cooperate. Gloria Jean Wiseman of Los Angeles made tape recordings and kept a written log of the barking. She then went to a city animal control hearing. She found that the tapes weighed in her favor.

Hearings don't always produce the results you want, however. Remember that the stakes are high. Because the dog may be ordered out of its home (in other words, destroyed), the hearing examiners take great care in making their decisions. One Los Angeles complainant was carefully cross-examined, and his evidence was thoroughly scrutinized. As a result, he was accused of splicing audiotapes to exaggerate the intensity of the barking. Another hearing ruled in favor of the dogs, because the complainant couldn't prove which dogs were barking. And like any court system, animal-control courts are backed up with cases. You may wait up to a year for a ruling, which might then get further dragged out with appeals.

That's what San Francisco's Jean Dunn found. She fought the noise from the dog next door for a year and a half. In the end, constructing a walkway that contained the dog away from her property line—and her bedroom—solved her problem. But the solution was hard won, and her neighbors fought her at every step. When the court ruled that the dog be fitted with an electronic collar that shocked it when it barked, the owners then accused Dunn of inciting the dog to bark.

The Neighbor's Pets

The kind of *extreme* violence against dogs that we mentioned earlier is unusual, but it does happen. It's an unconscionable deed, and when you think for a second about it, a dog can't help barking. It's the owners who are irresponsible (not that we advocate harming them either). But barking dogs sometimes cause neighbors to come to blows. A California woman who had suffered barking for two years snapped one day and shouted at the dog's owner. The dog owner's sister attacked her and hit her with a pail full of water mixed with dog feces. The pail cut her, and the wound required ten stitches. A judge subsequently found in the woman's favor and ordered the dog owner to pay $1,500 in damages.

Dogs bark because it's in their nature. And it's our nature to get irritated by it. Dogs bark when they're lonely, when they're bored, or when they're excited. Some of them will also howl when there's a fire engine that's racing with its off-key siren blaring. And they'll bark, of course, when they're angry. Fortunately, you can be sure they won't make any noise when they sleep, so you're guaranteed a couple of hours of silence now and then. In the meantime, look for allies in your neighborhood. Who will those supporters be? Most likely anybody and everybody who doesn't have a dog. And those who have quiet, well-trained pets.

Puppies can be trained by their owners not to bark, but not by frustrated neighbors. The dogs simply don't care what *you* think; they serve their masters. If you can get the owner to agree with you and train the dog, maybe you'll have some peace. Slip the phone number for several animal trainers under your neighbor's door. Or offer to share the cost for dog obedience training, if necessary. If the barking is driving you crazy, that's probably the only way to deal with the problem and still be friends (or at least not enemies) with your neighbors. If a barking dog is bothering you, in all likelihood your neighbor is aware that his dog is a bit loud, so your offer to have the dog trained might be appreciated. Try it; because it sure beats dog court.

Whether they are barkers or not, dogs should not be allowed to roam free. One of the most frustrating and annoying things about dogs on the loose is their "calling cards" (a nice way of saying their crap). You find it smeared on your shoes. Around your roses. Tracked onto your sidewalks. Yet, often when you complain, the pet owner says, "Hey, he has to use the bathroom somewhere!" "Sure," you reply, "but not in my yard." What should be done? He has to go somewhere, but preferably, where his master is right behind him with a pooper-scooper.

But if you can't get the master to walk the dog and clean up after it, one option is to make an impenetrable fortress of your yard, as one upstate New York family did. They moved into a tiny house in a 1930s subdivision at what had been the outer edges of Albany in the 1970s. Their house had only a wee, shaded backyard, but a nice, large, sunny front yard. So they decided to turn the front yard into a garden.

At first, they dug a few flowerbeds around the perimeter. But the impact of the neighborhood dogs, which were supposed to be tethered, was awful. They dug and deposited. In response, the homeowners decided to put up a nice, low (lower than city regulations for maximum height of a front-yard fence) white picket fence. When it went up, there was a hue and cry from all of the older residents that they were ruining the neighborhood.

Unfortunately, this resident eliminated one neighbor problem—the wandering dogs—and got another—neighbors complaining about the fence. At first blush, you might think that everybody (again, everybody except dog owners) would be repulsed by dog feces.

Some people accept dog doo, along with litter, as an inescapable reality of urban and suburban life, not as a preventable hazard. Generally, these people don't have gardens that are being defaced by dog feces, and they don't have children whose small stature puts them in close proximity to dog doo.

A fence is a reasonable, minimal, tool against animal toilet habits. And some sort of fence is probably legal in your community, especially if you can demonstrate that you have a dog problem. The most expensive stone wall or the least expensive chicken wire will work. For your neighbors' sake, if you do erect an anti-dog fence, keep aesthetics in mind. The more similar your fence is to other fences in the neighborhood, and the smaller it is, the better you will get along with all your neighbors. How small? That depends on how well the dogs in your neighborhood jump. Some dogs will jump over fences; others will just bypass them.

Invisible electric fences are also marvelous inventions. They are relatively inexpensive and can quickly teach an offending animal where its world ends and yours begins. These fences work well for *your* dog, not your *neighbor's* dog, because electric fences do nothing to keep out a dog that isn't wearing a special collar. One veterinarian said, "My personal experience is that they work extremely well when properly installed and the animal's properly trained." For dogs, these fences work best when the animal first encounters the fence as a puppy. Some dogs (remember that they're not all smart) will go charging through the invisible fence, then not be able to return to the house. So, if you do erect one, include a *visible* reminder that the *invisible* fence is there.

Every so often, depending on the phase of the moon, you may also be awakened in the middle of the night by the lovely, heartwarming sound of cats in heat. Yowling at a volume that's almost too high to measure, at a pitch that's nearly enough to shatter glass, cats call to each other.

A major problem dealing with cats in heat is that at 3 A.M. it's nearly impossible to determine whose cat is engaging in the mating ritual. If your neighborhood has six tabbies (that you know of), and one of them is auditioning for the Metropolitan Opera, who knows to whom the cat belongs? Even if you did know, no cat owner is going to keep his cat in for those few nights a year that it's in heat, mostly because it's not possible to predict with any certainty what nights are going to be terrors.

My advice, then, is to deal with the cat the first moment you hear it. Have a water pistol handy (add a few drops of ammonia to the water), find the cat, aim, and fire. Do this *before* the cat has become comfortable for the evening. Bring along a bright flashlight or flashgun and frighten the cat. No harm will come to the animal, and, with the exception of these fifteen minutes, you'll get a good night's sleep. The next day, you might mention to your cat-loving neighbors that there was a cat singing in heat all night long—did they hear it? A little guilt transfer will earn you points next time you need to ask your neighbor for a favor.

Perhaps the most significant way to calm a noisy cat is to have it spayed (females) or neutered (males). When cats lose their sexual drive, they also cease to make a ruckus about sex. Spayed and neutered cats are quiet cats. Indeed, some localities require cat owners to spay and neuter their pets, unless they have a breeder's license.

If there's a neutering law in your town, mention it to your neighbor. Should your neighbor think it's cruel or wrong or too expensive to have his cat neutered, then let animal control deliver the message to him. If your town doesn't have a spaying and neutering law, now's the time to work to get one passed. There's no good reason why you should have to suffer through a night of howling cats.

Cats wander, too, and while they're generally not as big a problem as dogs, they can easily elevate themselves to the status of nuisance. Here's what one enterprising Nebraskan did about a cat problem:

> One of my neighbors (a retired widow who waters her lawn at high noon when the temperature is a hundred degrees and "boils" my plants next to her yard in the process) obtained a "mature" cat a couple of years ago. Well, several mornings last spring I saw Ms. Kitty scratching in my newly tilled garden and leaving me a present right where my carrots were to be planted (almost the same spot every day). So I just put up my "hot wire" fence around the perimeter of the veggie garden, and that took care of that little problem. Ms. Kitty doesn't come near my garden anymore, even though she still comes in the yard, which is okay.

Electric fences don't harm animals. The shock they deliver is weak, but strong enough to teach an important lesson. And there are invisible solutions to dog and cat problems. For example, you can buy a number of pet repellents on the market, many of which are available in gardening catalogs.

But the responsibility for restraining a pet remains with the owners, right? You shouldn't really have to build a fence until you've exhausted your creativity in getting the owners to control it. And you can't fence everything. You can try talking to the owners, but be prepared to hear, "Penning an animal in is cruel." They didn't think Rover was a problem. Their dogs don't dig; their cats don't kill. The owners try to make you feel like an animal hater just because you want to be able to walk on your lawn without ruining your shoes. They try to make you believe that they have no alternative but to let their pets roam.

DIVERTING DOGS

- Use repellent sprays to keep them off your property.
- Fence your yard to keep your dog from wandering.
- Provide plenty of chew toys.
- Talk to the owners of a misbehaving dog before you resort to more drastic action.

Sometimes you have to take an aggressive, yet clever, posture against dogs. Dog owners usually don't get close enough to their pets' feces to appreciate how obnoxious it is. If a dog has inadvertently left its feces in your yard, well, you should return the dog doo to the rightful owner. Here's how one Washington State resident dealt with pooping dogs:

> One of our neighbors had an enormous dog that left enormous piles in several of our yards. We got together and solved the problem by putting the resultant piles in small cardboard boxes, which we gift-wrapped. We have rural mailboxes set in rows here so it was easy to set the packages in the neighbor's mailbox on a regular basis. It wasn't long before the offending dog was put on a leash.

Putting the poop through the dog owner's mailbox, on the owner's door mat, on the dog owner's newspaper—these are all ideas that have been used thousands of times a year by people fed up with having their yards become toilets. Returning the dog poop is one of the first strategies that angry neighbors employ. It meets with mixed results, depending on that ultimate wild card—human nature. One reason that merely dropping poop off where it came from sometimes doesn't work is that neighbors who let their dogs wander usually don't care in the first place about civility, or they would keep their dogs leashed. A poop now and then isn't a big deal, especially if only one neighbor—you—is involved. But a gang of neighbors can make a much stronger point. And it's also a lot more poop.

The Neighbor's Pets

There's also the problem of dogs peeing on your property. All dogs will scent-mark an object, even over another dog's marking. Scent-marking makes the area familiar to the dog and assists it in communicating, telling other dogs who has been there. Dogs have keenly developed senses of smell because they have an organ of smell in the roof of their mouths in addition to their noses. The olfactory area of the dog's brain is larger than that in a human's, too. When dogs dig up the ground where they have just urinated or defecated, they release scents into the air in addition to giving a visual sign to the visit. Although we don't understand why, some dogs may roll in foul-smelling messes, and that in itself is a good reason to keep them confined in the yard.

Dog walkers can also be a nuisance. For some reason, some dog owners regard your front lawn as a toilet for their dog. You've seen this at one time. Someone walking their dog while reading the newspaper, just letting the dog go in its favorite yard—and it takes considerable nerve to do that! If somebody's dog has selected your yard as its favorite, there are several steps you can take. First, talk firmly with the owner. Hang out on the porch—or stand stealthily at the front window—and confront the person. Most of the time, that works. If talking doesn't work, though, try yelling next time. If that doesn't work, take a photograph of the event and tell the owner that next time you see his dog on your lawn you'll deliver the photo to the police.

Creativity also counts in battling this kind of dog problem. Some solutions to your neighbor problems involve yelling at your neighbor, calling the police, filing complaints with the appropriate department—all the regular stuff. Health departments take an aggressive stand against people who violate leash and dog-poop laws—if they catch them.

It is the really inventive people who get results. Here's one inspired solution. A Cincinnati woman, who had tried everything to keep free-running dogs from "littering" her little patch of garden, finally took several of those liter-sized soda bottles filled with water, with a long cord attached around the bottle necks, and lay them in her garden. She claimed that it worked to keep the dogs out, so all of her neighbors followed suit. The dogs got tangled in the cord. Unfortunately, the dog owners were so incensed that their dogs returned home unrelieved that they went and crushed all of the bottles.

Dog owners rarely acknowledge that their dogs are a problem. (They're even less honest about the menace they create than are the parents of small children.) Here's one situation, from Illinois, that's typical:

My neighbor with the barking, messing beagle rarely heard the dog. But I did, right through my walls. I asked to have the poop scooped from my yard, so I could mow. She replied by telling me that she never heard her dog howl, that it only pooped in her yard (never mine), and that she didn't let it run loose every day, only three to four times a week! I finally erected a fence, at great cost to me. Three months later, she got rid of the dog.

Denial is the first defense of dog owners. If your neighbors refuse to believe their pet is raising a ruckus, get some proof. Take some pictures and show the owner. Catch the dog in action: an indelicate pose, a garbage pose, a fight pose. If you can't get satisfaction from the neighbor, go public. That's what worked for one Baltimore, Maryland man. He took his fight public by posting signs with the dog's photo. The sign said:

WARNING: *This dog has been seen unleashed, wandering the neighborhood, pooping on people's yards. If you see this dog, please call Animal Control. Dog feces spread disease, and are especially dangerous to children.*

Everyone in the neighborhood knew who the dog was. The embarrassed neighbor collected her dog, Juniper, and kept him penned after that.

Roaming dogs are also dangerous. While it's not the purpose of this book to delve into dog psychology, it's important to point out a few aspects of how dogs behave. All dogs have a boundary zone. When a threatening person or animal enters that zone, the dog will either run away or attack. Which action it chooses depends on the breed of dog and how brave that particular dog is. The size of this defense zone also varies. There's no way to know ahead of time what a dog will do when a stranger approaches. A small child who walks over to a roaming dog, intending to pet the animal, could succeed in his mission or could get his hand bitten. An untrained dog or one whose training has not been maintained is more likely to attack. But there's no way to tell by looking at the dog what its obedience level is. Dogs respond aggressively to a wide range of activities. Running and yelling—things that children do all the time outside—can provoke a sudden, violent attack.

Virtually all localities recognize the potential danger of roaming dogs. Arizona State University enacted a strict leash regulation after a pack attacked three Seeing Eye dogs. When blind students on campus asked other students to keep their dogs leashed, they were met with shouts. Penalties for letting your dog off its leash (which can't be longer than six feet) include stiff fines and having the dog impounded. In Tempe, Arizona, where Arizona State University is located, the *criminal* penalty for letting your dog roam free can be up to $2,500, plus six months in jail. That's a law with more bite than bark!

Pasadena, California, has a similar law. But in that city you can also be fined for letting an unleashed dog run around your own unfenced yard. Why? Because a dog can suddenly change its mind and run into the next yard where there are small children. (Or dash across the street and get hit by a car.) Do leash laws really improve neighborly relations? The answer is yes. In the first year that Orange County, California's leash law was in effect the complaints against dog owners dropped 15 percent; euthanasia of dogs dropped by 30 percent.

Pets suffer, too, when their owners let them roam. One day King doesn't come home, and the next the owner gets a call that he's been killed out on the highway. Eventually, that's what happened to Juniper. The owner let the dog roam just once more, and during that once more a car hit the dog. Sometimes the non-dog-owning neighbors get twice suckered. Once when the roaming pet digs and dirties. Then when they witness the pet being hit by a car, rush the poor animal to the vet, and end up stuck with the bill. Louise Anderson of Flagstaff, Arizona, had this story:

> A stranger's cat cost me fourteen hundred dollars, and it had to be put to sleep after a valiant effort to save it. I just finished paying off the bill. The cat's owner (whom I did manage to find) didn't offer a penny toward it, not even the "final cure."

What this story shows is how insensitive some pet owners can be, not only to their neighbors, but to their pets as well. A dog owner who lets his dog wander freely or a cat owner who never or rarely lets the cat inside may not care about the animal at all. In these cases, there's little chance that you'll convince the owner to restrain the pet.

Roaming dogs can do terrible things. Every dog is a "nice doggy" in the eyes of its owner, but dogs are unpredictable and can strike with sudden severity. Here's what one mother said about what happened to her child:

> My son underwent three and a half hours of emergency surgery the night of his accident, spent four days in hospital being fed by IV, had approximately three to four hundred stitches to his face, forehead, septum (inside his nose), outside bridge of his nose, under his nose, inside his upper lip, and a skin graft done on his cheek.

Here's an even worse tragedy. In 1988, in St. Petersburg, Florida, a dog bit a six-year-old child. The child got an infection from the dog bite and died.

CAT AND DOG DISEASES

There are numerous diseases that humans can catch from cats and dogs. The most serious is toxoplasmosis, a parasitic organism that's found in cat feces. In adults, toxoplasmosis causes flu-like symptoms. A pregnant woman who is exposed to toxoplasmosis may pass the parasite to her unborn child. Depending on how far along she is, there is a risk of miscarriage or stillbirth, or the baby may suffer serious birth defects.

Dogs sometimes carry hyatid disease, a worm-borne infection. These worms form eggs that produce giant cysts—the largest known contained twenty gallons of fluid—in bones and organs. The only way to remove them is through surgery.

Ringworm, a fungal infection, can be caught from both cats and dogs, although you're more likely to contract it from a cat. Ringworm produces severe itching; it can also be passed from person to person. Cat bites can produce a variety of infections, and all cat bites require a doctor's attention.

Children sometimes get toxocariasis, a worm infection, most often from eating dirt contaminated with dog feces. In humans the toxocariasis worm penetrates the intestinal wall, enters the bloodstream, and then finds its way to the liver and lungs. In rare instances, toxocariasis causes blindness. Children are most at risk, because they often play in proximity to dog feces. Hookworms and roundworms can also be transmitted from dogs to people. Dog feces are also dangerous, not to mention disgusting, especially when you discover a pile in your yard—and you don't have a dog. Rats enjoy dining on dog poop, so the neighbor who lets his dog roam may be helping feed a rat family. If you walk barefoot in your yard and step on dog poop, you run the risk of becoming infected with these parasites.

Leptospirosis is another disease that people can catch from dogs. It can produce jaundice or meningitis. Salmonella is always a danger from any animal's feces; it causes fever, diarrhea, and vomiting. Young children, the elderly, and those with weakened immune systems are most at risk.

Wandering dogs can pick up tick-borne diseases, including, and most dangerously, Lyme disease and Rocky Mountain spotted fever. Rabid animals can also bite wandering dogs, helping to spread that fatal disease. Not everyone has their dog vaccinated against rabies, although all laws about it are strict.

WHO TO CALL ABOUT ROAMING PETS

What are the laws regarding roaming dogs and what they do to other animals? They vary from region to region, but this Georgia statute is pretty typical:

> The owner, or, if no owner can be found, the custodian exercising care and control over any dog which goes upon the land of another and causes injury, death or damage directly or indirectly to any domestic animal which is normally described as livestock or fowl, shall be civilly liable to the owner of the domestic animal or fowl, for damages, death or injury caused by the dog. The liability of the owner or

custodian of the dog shall include consequential damages. The provisions in this section are to be considered cumulative of other remedies provided by law.

In other words, it can be expensive for a dog owner to let his dog roam freely.

But what if you don't want to spend five hundred dollars or more to build a fence? What if your neighbor simply won't keep his dogs penned in? What if your neighbor doesn't want to bother with a leash? After all, it's much easier to let two golden retrievers walk by themselves, rather than try to restrain them on a leash.

When people get whopping mad, mad enough to fight—or worse—they often end up losing. His neighbor shot actor Jameson Parker when one of his four dogs urinated on the man's lawn. For months, Parker's dogs had been leaving calling cards in neighbors' yards, and they had been complaining loudly. The trigger-happy neighbor was charged with attempted murder and assault with a firearm; it didn't help that Parker, his wife, and the dogs moved out of the neighborhood, because the neighbor now resides in prison.

The police aren't going to be much help. What are they going to do? Arrest the dog? Dogs don't like being put in police cars, and policemen don't like having to try to get them in. They're generally too involved with more important crimes to worry about wandering dogs. Dealing with dog problems is time consuming: First you have to collar the dog; then read the dog's collar. Then match the information on the collar with a name and address; then find the owner. In any event, there is little police can do. Animal control or the health department is the avenue that's going to be most productive. Animal control is charged with the responsibility of dealing with stray and wandering animals. They're experienced and they know how the animals—and the owners—behave.

Pet poop can be a problem even when it remains on the owner's property. It all depends on the pet. Gus, a Vietnamese potbellied pig, like many others around the country, got a lot of attention. He's trendy, he's cute (humble, too), and he's expensive. Yet the Humane Society of the United States says pigs shouldn't be kept as pets because they need to wallow and root—messy, unsanitary behavior. Many local laws, therefore, ban farm animals from residential areas. (Though from Gus's point of view, the ban on residential pigs in Orlando, Florida, was the problem.) Owners of the little pigs will counter that at thirty-five pounds, they're smaller than many large dogs. And smarter, too.

Maybe no one would have said a word about Gus if it hadn't been for the small waste disposal problem. It's not that the animal didn't have tidy toilet habits; it's just that the owners had a little trouble disposing of the waste. Gus *was* litter trained, meaning it buried its waste, but pigs produce a lot of poop. One pig owner might have to dispose of twelve to fourteen hundred-pound cans of waste a week!

Why mention potbellied pigs? If you live downwind of a neighbor who's thinking of getting one—assuming you can find out ahead of time—be forewarned. You can stave off a pig problem by trying to talk your neighbor out of getting this pet. It's possible that they aren't aware that potbellied pigs are so odoriferous; the literature promoting them doesn't always point that out.

In addition to their calling cards, roaming pets leave other reminders of their visits. Dogs tear up your garbage. So do cats, but they're daintier about it. Most of all, roaming animals harm other animals. If you feed the birds, you know what a problem dogs and especially cats can be in your backyard.

Lobby Against Pet Problems

Have roaming dogs, aggressive cats, potbellied pigs got you down? Are these animals pawing at your garden, terrorizing your children, making your bird feeder look like a saloon after last call? Then consider lobbying to change your local animal control ordinances. Here are some legislative objectives:

- Although most municipalities allow cats to roam freely, the Lewisville, Texas, animal control supervisor points out that laws requiring that cats be kept on their owners' property reduces rabies risk.
- Limits on the number of pets that any one household can have are another strategy. It's easy to be a good dog owner; it's much harder to be a good owner of four dogs.
- Propose harsher penalties for unleashed dogs. Posting of signs pointing out those penalties can go a long way toward solving roaming-dog dangers.
- Impose stringent penalties for dog bites. There should be no free bites; penalties should increase for each subsequent offense.
- Establish enclosed dog runs, so that dogs can run around without a leash and without endangering children.
- Enact pooper-scooper laws.
- Enact noise ordinances that limit the decibel level that dogs can produce, either all day or during sleeping hours.
- Permit only those cats that have been vaccinated and sterilized to roam freely.
- Require that ferrets be registered and vaccinated against rabies.
- Restrict potbellied pigs to non-urban areas.
- Clearly define what a farm animal is, and ban farm animals from residential areas.
- Forbid the keeping of exotic pets such as potbellied pigs, ferrets, or raccoons.

The Neighbor's Pets

Most changes in animal laws are implemented because concerned or irate citizens are bothered by animals. Roaming dogs present an easier legal fight than roaming cats. Cats are considered feral animals, in other words, it is their nature to wander. Interpretation of the law is left to judges and juries, so the outcome of a lawsuit isn't always predictable. Here's one story from New York. It demonstrates that cats don't just go after birds in the cartoons; it happens in your yard, too. But if your yard is in a wildlife sanctuary, your cat had better not be the killer.

The Clarks of Gates Mills, New York, wanted the killing in their yard to stop. It was their neighbor's cat that was doing the dirty deeds. A black cat belonging to a couple by the name of Hurley was visiting the Clarks' property and killing birds. Hunting, trapping, and roaming animals are forbidden in the village, and police charged the Hurleys for letting their cat roam. The Clarks weren't cat haters. They kept their own cat on a harness when it was outdoors. But that's cruel, countered the Hurleys. The Clarks were serious friends of the birds. Melanie Clark was a volunteer rehabilitator for the Lake County Metroparks, and she took home sick and injured birds to nurse them back to health. She tracked the movements of hawks and other species she saw in her yard and sent the information off to Cornell University.

The Clarks kept bird feeders that attracted several thousand birds a year. And, as they liked to point out, household cats cause at least 20 percent of the injuries at the Metroparks Wildlife Center where Melanie Clark volunteered.

DEALING WITH WANDERING CATS

- Make sure you take action against the pet's owner, never the pet.
- Better yet, take pictures and get evidence of the offenses.
- Start a poop collection for the pet's owner. Drop off your gift when the neighbors are entertaining.
- Get in touch with your neighbors who are gardeners. They're likely not to be too keen on a neighbor's pet pooping in their yard.
- Get in touch with the parents of young children. Animal excrement and kids don't get along. You'll find allies among parents.

TOTAL CRITTER CONTROL

In their defense, the Hurleys admitted that although a black cat might have been killing birds at the Clarks' feeder, it was not their cat. Like most pet owners, they said, "Not my animal." They said she never left their yard. The Clarks insisted it was not so. The matter went before a judge, who decided in favor of the Clarks.

Sometimes you wouldn't mind one dog or cat, but some people just can't stop once they start collecting anything. Here's what happened to one St. Petersburg, Florida, homeowner:

> First let me say that I own a cat; it's a mouser. I have a neighbor two doors down who has about thirty cats! There are several howling catfights every night, and these other cats steal my cat's food. They were constantly spraying everywhere, cat crap all over our lawns and occasionally even on the hoods of our cars, on the stairways, and even on my neighbor's front-porch bar! And that bar is about twenty-three feet off the ground! (We live in stilt houses in a beach community.) After explaining things to the cat people several times and having animal control come out weekly to trap the critters, I went out and bought a Havahart raccoon trap. We were averaging 1.5 cats per day and were getting to know the entire staff at animal control on a first-name basis, when we found out the guy was coming onto our property and taking them out. The cat man then kept all these felines in the house with him, that is until they could no longer stand it. More trapping ensued. Last week the cat man built a cage about 4' x 6' x 8' in his backyard and stuffed as many of them as he could in it. He still sneaks onto our property whenever one of the @!#$%! things are missing.

Sometimes when the animal control officials won't (or can't) act, you have to do what you can on your own. And people with thirty—or even just five—pets are, well, a bit eccentric. They'd sooner cut off their right arms than part with a single animal, so don't even bother asking. (If you don't think they're a bit kooky, just look at their wills.) In many less-than-urban localities, local laws may permit people to have as many pets as they wish.

The only way to thwart these multiple-animal owners is to try another approach. Are the animals a health threat? If so, see what your local health department can do. Are the animals noisy? Then local noise regulations may be violated. Is your neighbor breeding animals? If so, he or she may need a license. Are the animals being treated cruelly? Contact your local humane society. Animal control is only one of many potential agencies that can bat in your favor. Government bureaucracy is so varied and so large that if you explore a little, you can find the appropriate agency that you need (with a bit of luck). If the animals seriously bother you, you need to move, see an allergist, or soundproof your house. Oh yes, and plant lots of fragrant flowers.

The following story seems like a bad joke. A family in a Denver suburb had 105 poodles, although local ordinance permitted only three dogs. When someone turned in the owners, who were dog breeders, they faced a choice: Move or get rid of the dogs. They took a third option: They got a dog-law lawyer and fought. (The lawyer had once represented a Chicago-area couple with 140 poodles, speaking of overspecialization in the law.) The couple had moved into their home in the early 1960s to engage in poodle breeding, before the law was enacted; only later was the area annexed into a suburban town. No one complained until 1991. The judge let the poodles stay.

The moral? Even in what seems like the most outrageous case, you can't count on achieving victory through the law. This case also speaks to a general principle when it comes to dealing with annoying neighbor pet tricks: Act swiftly. You don't have to take instant, irrevocable action against the neighbor with the annoying pet. *But you do have to make your opposition clear from the moment you learn about the animal.* Time favors the status quo. (This general observation is true for other annoying tricks that neighbors perform, too.) If the pets were in place before you appeared, you may simply be stuck with them.

Paul and Martha Rogers wanted to enjoy their retirement in a country setting. Their neighbors, who were there first, said they just wanted to make a living. The problem was, they operated a dog kennel. Of course, the Rogers fought with a lawsuit. Kennels are always controversial; they're noisy and sometimes even smelly. As with other controversies, the themes in this fight were familiar. The Rogers said they had to stuff cotton in their ears to get any sleep. The Van Vorsts, the people who owned the dogs, pointed out they were there first and operated a business that had never caused any complaints in seventeen years. The county even gave them a permit to expand their operation to keep seventy-five dogs, doubling their capacity. That's when the Rogers family first started to complain.

Eventually, the Rogers family lost the case. Keep this important point in mind: If you are considering moving into a neighborhood with an animal problem, don't count on being able to eliminate that problem once you've moved in. The odds are not necessarily in your favor.

In Ojai, California, the problem was a little more exotic: birds. Not just little sparrows at a feeder. Major birds. Parrots. Loud birds. Bird breeder Kay Nesbit had an aviary, the Bird in the Hand Exotic Bird Farm, and it had the neighbors squawking. Nesbit kept more than one hundred parrots, cockatoos, macaws, and other exotic birds.

TOTAL CRITTER CONTROL

She said she was doing good work, breeding birds, some of them listed by the federal government as endangered species, and shipping them across the United States. Without her birds, she said, people would be buying birds caught in the wild, a practice that further depletes the population. So she had many supporters, but none of them were among her neighbors, who complained about the screeching and squawking. Neighbors said the sounds invaded their homes. Nesbit said the neighbors were hearing peacocks on their property and birds attracted to her home because of the parrots. The fight went on for years. There were petition drives, public testimony, a slander suit filed by Nesbit, and then county-ordered sound modifications.

Nesbit didn't understand why neighbors complained. She lived in a rural area zoned for agricultural uses. Schoolchildren visited to learn about ecology, and she took her birds to visit facilities for the disabled. At last report, Nesbit had built a five-thousand-dollar sound barrier, but her neighbors claimed they couldn't tell the difference. Neighbors separated from Nesbit by as much as five acres still found the noise annoying. Measurements of the sound found the noise was within acceptable county guidelines. To Nesbit's psychic benefit, her closest neighbor supported her. At a hearing, the neighbor said she found the sounds to be "natural," if sometimes "quite loud."

In Clearwater, Florida, a green-wing macaw was the problem. This bird was so loud and aggressive, said one neighbor, that it was "almost like a Doberman pinscher that was going to attack you."

In another animal case, police surprised riders from Glenview Farms with summonses after neighbors complained about the noisy horses. The farm, located in Long Island's Laurel Hollow, is a well-known stable of the riding set. It seems the horses violated a 6 P.M. to 8 A.M. noise restriction in the village. Glenview Farms, where they also gave riding instruction, came under censure as well. The *clip-clop* of horses and the riding instructors' commands bothered some neighbors. The police came but didn't arrest anyone.

But there are people who had it worse than the Rogers clan. What about roaming farm animals? Robert and Nancy Disson were gardening in their front yard when a goat attacked them. They lived near a house that some neighbors had turned into a farm.

Another neighbor had hung double windows to block out the barnyard noise. Everyone in the area was used to the occasional stray chicken. But a goat? It was an upscale neighborhood, although the barnyard was there first—since 1967. Surrounded by homes selling for $175,000 to $200,000 was a farm belonging to Barnabay and Althia Wittaker, an elderly couple who had kept chickens, ducks, sheep, goats, and horses before there was a subdivision. In 1990, the neighbors filed a lawsuit against the Wittakers.

The Neighbor's Pets

The neighbors had a lot to complain about. Obviously, the farm reduced property values; it was poorly kept and relatively unattractive. The odor that emanated from it was a little stronger than the smells that came from the other neighbors' ovens. It attracted flies and other insects. The fowl didn't stay put and often defecated on neighbors' cars. And then there was the noise. The Wittakers spent thousands of dollars for their legal defense—and won.

While this book is about *winning* a war, it's important to point out that not all wars are going to be won. It's important to know this ahead of time. So, what do you do when the law is on the side of the animal owners? You take frequent vacations, use lots of perfume, move your air-conditioning condensers to the other side of the house, and buy double-thick windows. But mostly you wait.

Just because the law allows people to keep animals, it doesn't necessarily allow their animals to be a nuisance. If a goat attacks you, bring that matter to somebody's attention. If a wandering dog trespasses and frightens your children, call animal control. If a chicken uses your newly waxed car as a toilet, make a stink in court.

You may lose the war over animals, but you and your neighbors can also win enough battles to make it very tiresome for the animal owners. Wear them out. Use guerrilla tactics and attrition. One small-claims court case every now and then isn't going to matter. But if a dozen neighbors file suits twice a year, the animal-owning neighbors may decide it's not worth the trouble to keep all of those critters.

EXOTIC PETS

What happens when your neighbor collects poisonous snakes? What if you think the neighbor's pets are dangerous? Often there's not much you can do unless a local law is on your side. People who own fierce animals are religiously devoted to them. They are often fierce themselves. These people don't care what *you* want. They don't even care what every *neighbor* thinks. They want their pets, if you can call them that.

What if you lived next to these guys? Two brothers near Ottawa, Canada's capital city, had their neighbor a little riled over their pets of choice: cougars and wolves. The wolves and cougars were smelly, noisy (the wolves howled at night), and worst of all, as the fairy tale goes, they had very, very sharp teeth.

The two brothers and their menagerie ended up outside Ottawa, because they had had to leave their previous residence in Waterloo (northwest of Toronto) quickly after one of the cougars clawed a boy. No matter what, the brothers were determined to keep their six pets (three of each). If the neighbors were eventually successful in forcing the wild animals out, one brother threatened revenge. "I'm going to bring in pit bulls and Rottweilers and paint everything fluorescent orange," he bellowed.

Neighbors like this are as dangerous as their pets. The safest way to deal with them is to ignore them. Teach your kids not to play anywhere near these animals, and keep an eye on your children at all times because they'll try to do it anyway. And then keep protection close at hand; a cayenne-pepper-based spray made for protection against bears is one option. If your neighbor keeps really dangerous pets, and you're worried that they might get loose, then you might consider getting a license and a gun and learning how to use it. Putting up the "For Sale" sign might be another option.

Strange neighbors with dangerous pets have to cooperate in some small ways, so your neighborhood won't be, well, a complete zoo. These neighbors still have to obey the law, and most cities and towns forbid keeping wild animals as pets. Even those that allow such animals in private homes still have nuisance laws to keep noise and odors to a minimum. When you smell or hear something unpleasant, call the owners (keep the number by your phone). That may calm things down for a little while.

In Warrenton, Virginia, the problem was pit bull terriers. That's pit bulls in the extreme plural. How would you like to have three dozen pit bulls living next door? That's the number Jewel Metta owned. Imagine trying to barbecue with thirty-six pit bulls growling and barking as you turn the burgers. Metta's neighbors called the local animal control board. Health officials. Zoning officials. Trouble was, there was no violation. What? No law broken? Hard to believe, but it was true. When questioned about the complaints, Metta said she didn't regard her dogs as "especially fierce."

Keeping dangerous animals isn't necessarily against the law. But if they bite or kill somebody—that's against the law. In the meantime, do the best you can and protect yourself. If you feel that an animal is a menace, you are right. The definition of a dangerous animal is one who makes people feel afraid.

Sometimes people moan and groan about threatening animals, but they never actually do anything about a dangerous animal in the neighborhood until it's much too late, usually after some sort of casualty. In one case, even though neighbors thought two dangerous dogs were roaming the neighborhood, no one took action. And another dog died as a result.

Sometimes a name is destiny. Taco, a Dade City, Florida, chihuahua, was killed by two larger dogs roaming the neighborhood. His owner had released the tiny dog in his yard, a presumably safe place, and had gone back indoors, when a yelp alerted her to trouble. She saw one of the large dogs carry away the small animal. Although neighbors had complained, rather gently, about the roaming large dogs, nothing had ever been done to control them. Animal control told neighbors to file a civil suit, something that didn't interest them. The killer dogs' owners were charged with keeping vicious animals.

It could have been worse—these dogs could have killed a child. To the owners, however, the dogs were as gentle as pussycats. That's usually the case: Lost dogs treat their *owners* well. "He won't bite." That's what they all say. Not infrequently dog owners just don't care if their dog is vicious. Sometimes the trespass and damage are habitual. And someone finally takes notice too late.

Sue Mehl in Georgia's Polk County lost her poodle when the German shepherd next door came into her yard, killed the smaller dog, and stole its bone. It wasn't the first time the shepherd had trespassed; in the past the dog had destroyed two pairs of Mehl's shoes, five outside doormats, and had repeatedly overturned garbage cans and strewn refuse all over Mehl's lawn. Each time, Mehl visited her neighbor, Al Penney, who did nothing to restrain his dog. In the end, the two wound up in court with Penney paying Mehl two hundred dollars in damages.

Merely complaining doesn't always produce results. But don't let this story keep you from trying. In Ohio, Felix Fleming asked neighbor Richard Archer to

CONTROLLING BAD NEIGHBORS AND BAD PETS

- Make your concerns known to the pet's owner. Show the owner newspaper articles in which the neighbors were awarded judgments in lawsuits.
- If that does nothing, talk with as many appropriate officials as you can. See if the animals are violating any laws—business, sanitary, noise—whatever you can uncover. If nothing else, your complaints make for a stronger case if you end up in court.
- Keep kids and vicious animals far apart.
- Get some protection in the form of pepper spray, or whatever you consider appropriate.
- Always remember: Any dog can kill or maim.
- If you can't persuade your neighbor to keep their animals inside, you might want to try a cat repellent spray. Get Off My Garden is designed to keep pets and wild animals away from your plants and patio.

refrain from walking his dogs while the Fleming children were waiting for the school bus. The dogs were habitually interested in knowing what the children smelled like, and neither the children nor their parents enjoyed that. The situation escalated into a classic neighbor war, with Archer accusing the Fleming children of teasing his dogs and Fleming accusing Archer of harboring vicious animals. The families wound up in court, and in the end, Archer moved.

Here's one last statistic that will be food for thought. More than half of all the households in North America have pets. That's a lot of dogs and cats, birds and potbellied pigs. It's impossible to tell who might move in next door with a disagreeable critter. But when it comes time for you to move to a new home, you might want to add "neighbors' pets" to your house inspection list.

RESOURCES

GROUPS

American Dog Owners Association
1654 Columbia Turnpike
Castleton, NY 12033
(518) 477–8469
www.adoa.org

Bat Conservation International
P.O. Box 162603
Austin, TX 78716–2603
www.batcon.org

Beaver Defenders
Unexpected Wildlife Refuge, Inc.
Box 765
Newfield, NJ 08344
(856) 697–3541
www.animalplace.org

Beavers: Wetlands and Wildlife
146 Van Dyke Road
Dolgeville, NY 13329
(518) 568–2077
www.beaversww.org

Fund for Animals National Headquarters
200 West 57th Street
New York, NY 10019
(212) 246–2096
www.fund.org

Zebra Mussel Information Clearinghouse
Susan Grace Moore, Librarian
State University of New York
Brockport, NY 14420
(716) 395–2638

MAIL ORDER CATALOGS

Audubon Workshop
5200 Schenley Place
Lawrenceburg, IN 47025
(812) 537–3583
www.audubonworkshop.com

Brookstone
Numerous locations across the United States.
1–800–926–7000
www.brookstone.com

Gardens Alive!
5100 Schenley Place
Lawrenceburg, IN 47025
(812) 537–8650
www.gardensalive.com

Jackson and Perkins
1 Rose Lane
Medford OR 97501
1–800–292–4769
www.jacksonandperkins.com

Resources

Peaceful Valley Farm Supply
P.O. Box 2209
Grass Valley, CA 95945
(916) 272–4769
www.groworganic.com

Plow and Hearth
P.O. Box 6000
Madison, VA 22727
1–800–494–7544
www.plowandhearth.com

Seventh Generation
212 Battery Street, Suite 4
Burlington, VT 05401–5281
(802) 658–3773
www.seventhgen.com

Solutions
P.O. Box 6878
Portland, OR 97228
1–800–342–9988

Sporty's
Clermont Airport
Batavia, OH 45103–9747
1 800 543 8633

PUBLISHERS & PUBLICATIONS

Bio Integral Resource Center
Box 7414
Berkeley, CA 94707
(415) 524–2567
www.igc.org/birc
Publishers of *Common Sense Pest Control* by Sheila Daar and Helga and William Olkowski

The National Coalition Against the Misuse of Pesticides
701 E Street, SE, Suite 200
Washington, D.C. 20003
(202) 543–5450
www.beyondpesticides.org
Publishers of *Safety at Home: A Guide to the Hazards of Lawn and Garden Pesticides and Safer Ways to Manage Pests*

www.greenchoice.com has a newsletter that outlines new pest control products that do not harm the environment. Subscriptions are available through the Web site.

www.keyed.com/birc/pubrep.htm is an excellent Web site with more than fifty publications on integrated pest management. The Bio-Integral Resource Center (see previous page) is a nonprofit, scientific, and educational organization dedicated to reducing the use of toxic pesticides. They also offer memberships, which include a subscription to their magazine and newsletter.

www.watoxic.org has a variety of publications that can be ordered online. Some of the topics include how do deal with fleas, moths, carpenter ants, spiders, and cockroaches. Their address is Washington Toxics Coalition, 4649 Sunnyside Avenue North, Suite 540, Seattle, WA 98103; telephone is (206) 632–1545.

www.epa.gov/OPPTpubs/citguide.pdf/ offers copies of the *Citizens' Guide to Pest Control and Pesticides Safety*. It can be ordered by telephone at 1-800-409-9198.

http://cals.uvm/edu/ctr/pubs/populr5d.htm is run by the University of Vermont and offers leaflets on pest management. They can be accessed through the Web site or purchased online.

North Dakota State University, through its Extension Service, offers more than seventy publications on aphid management, grasshopper biology, integrated pest management, mosquito control, blister beetles, and pesticide use and safety. Their Web site is www.ext.nodak.edu/extpubs/pests.html.

The Maryland Cooperative Extension offers twenty-five publications on the subject of pest control, including booklets on termites, pesticide sprayers, gypsy moth caterpillars, Lyme disease, and wood boring beetles. Their Web site is www.agnr.umd.edu/CES/Pubs/topics/pestcontr.html.

Resources

Pest Control magazine publishes an annual REDBOOK that lists their complete line of pest management products and services. This publication also includes helpful information about state regulatory offices, EPA offices, and related organizations and agencies by state. REDBOOK is available by calling 1–800–598–6008. Outside the U.S., call (218) 723–9180. Some of their specific publications, all available by direct mail, include *Truman's Scientific Guide to Pest Control Operation* (for insects and vertebrate pests), and *Vertebrate Pest Handbook* (good information on mice and rats). To order *Pest Control* magazine, call 1–888–527–7008, or order online at www.pestcontrolmag.com.

WEB SITES

http://pested.unl.edu/catmans/aganimal/aganimal.htm is a training manual used by the University of Nebraska-Lincoln. It is highly detailed and focuses on the identification and life cycles of insects.

www.co.broward.fl.us/ppi02100.htm is a site dedicated to "common sense pest control" that looks at less toxic ways of dealing with pests.

http://ipmwww.ncsu.edu/biocontrol/biocontrol.html explains what biological pest control is, and how to make it work for you.

www.futuregarden.com explains how to use natural predators to control pests.

www.doyourownpestcontrol.com includes an identification guide to household pests and a list of commonly asked questions.

www.asktrapperjohn.com is a site where you can ask the host (a published author on the subject) any questions relating to animal or pest control.

www.goodnet.com/~animals/UWS/ is not solely dedicated to pest control (there is also material about cruelty laws), but there is helpful information on humane ways of dealing with a variety of birds.

http://nppests.cas.psu.edu/mammal.htm is an excellent Web site with information on how to deal with bats, bears, geese, groundhogs, mice, moles, and more.

www.recyclenow.org/less-toxic/ deals with ants, cockroaches, spiders, and fleas in considerable detail.

http://www.uos.harvard.edu/ehs/hot_topics/ has information on many different insects and how to identify and deal with them in and around your home.

BOOKS

American Horticultural Society Pests and Diseases: The Complete Guide to Preventing, Identifying and Treating Plant Problems by Pippa Greenwood (DK Publishing, 2000).

Bug Busters: Poison Free Pest Controls for Your House and Garden by Bernice Lifton (Avery, 1991).

Bugs, Slugs & Other Thugs: Controlling Garden Pests Organically by Rhonda Massingham Hart (Storey Books, 1991).

Cougar Attacks! by Kathy Etling (The Lyons Press, 2001).

Common-Sense Pest Control by William Olkowski, Sheila Darr, and Helga Olkowski (Taunton, 1991).

Dead Daisies Make Me Crazy: Garden Solutions Without Chemical Pollution by Loren Nancarrow and Janet Hogan Taylor (Ten Speed Press, 2000).

Ecological Approach to Pest Management by David J. Horn (Guilford Press, 1988).

Dan's Practical Guide to Least Toxic Home Pest Control by Dan Stein (Hulogosi, L.L.C., 1991).

Dead Snails Leave No Trails: Natural Pest Control for Home and Garden by Loren Nancarrow and Janet Hogan Taylor (Ten Speed Press, 1990).

Deer-proofing Your Yard and Garden by Rhonda Massingham Hart (Storey Books, 1997).

Ecologically Based Pest Management: New Solutions for a New Century by the Committee on Pest and Pathogen Control (National Academy Press, 1996).

Resources

Entomology and Pest Management by Larry P. Pedigo (Prentice Hall, 2001).

A Guide to Lawn, Garden and Home Pest Control Products by W. T. Thomson (Thomson Publications, 1996).

Handbook of Pest Control: The Behavior, Life History, and Control of Household Pests by Arnold Mallis, Dan Moreland (editor) (Mallis Handbook & Technical Training, 1997).

The Homeowner's Pest Control Handbook by Gene B. Williams (Arco Publishing, 1978).

Insect Pest Management Techniques for Environmental Protection by Jack E. Rechcigl (editor), Nancy Rechcigl (editor) (Lewis Publishers, 1999).

Introduction to Insect Pest Management, 3rd Edition, by Robert Lee Metcalf (editor) and William H. Luckmann (editor) (Wiley-Interscience, 1994).

Least Toxic Home Pest Control by Dan Stein (Book Publishing Co., 1994).

Living With Wildlife: How to Enjoy, Cope with, and Protect North America's Wild Creatures Around Your Home and Theirs by Diana Landau, Shelley Stump (editor) (Sierra Club Books, 1994).

Natural Insect Control (21st Century Gardening Series, Handbook #139) by Warren Schultz (editor) (Brooklyn Botanic Gardens, 1995).

Natural Pest Control: Alternatives to Chemicals for the Home and Garden by Andrew Lopez (Chelsea Green Publishing, 1998).

The Organic Gardener's Handbook of Natural Insect and Disease Control: A Complete Problem-Solving Guide to Keeping Your Garden and Yard Healthy by Barbara W. Ellis (editor) (Rodale, 1996).

Organic Pest Control Handbook by Pest Publications Staff.

Outwitting Bears: Living in Bear Country by Gary Brown (The Lyons Press, 2001).

Outwitting Coyotes by John Trout, Jr. (The Lyons Press, 2001).

Outwitting Critters: A Surefire Manual for Confronting Devious Animals and Winning by Bill Adler, Jr. (The Lyons Press, 1997).

Outwitting Deer by Bill Adler, Jr. (The Lyons Press, 1999).

Outwitting Deer in Your Garden by Peter Loewer (The Lyons Press, 2002).

Outwitting Neighbors by Bill Adler, Jr. (The Lyons Press, 2000).

Outwitting Mice and Other Rodents by Bill Adler, Jr. (The Lyons Press, 2001).

Outwitting Squirrels by Bill Adler, Jr. (Chicago University Press, 1996).

Outwitting Ticks by Susan Carol Hauser (The Lyons Press, 2001).

PCT Field Guide for the Management of Structure-Infesting Ants by Stoy A. Hedges and Dan Moreland (editor) (GIE, Inc., 1995).

Peterson First Guide to Urban Wildlife by Sarah Landry (Houghton Mifflin, 1998).

Slug Bread and Beheaded Thistles: Amusing and Useful Techniques for Nontoxic House-keeping and Gardening by Ellen Sandbeck (Broadway Books, 2000).

Solving Coyote Problems by John Trout, Jr. (The Lyons Press, 2001).

Squirrel-proofing Your Home and Garden by Rhonda Massingham Hart (Workman Publishing, 1999).

Squirrel Wars: Backyard Wildlife Battles and How to Win Them by George H. Harrison, Kit Harrison (editor) (Willow Creek Press, 2000).

The Termite Report: A Guide for Homeowners and Home Buyers on Structural Pest Control by Donald V. Pearman (Pear Publishing, 1988).

Termites: Biology and Pest Management by M. J. Pearce (CABI Publishing, 1998).

Termites and Borers: A Homeowner's Guide to Detection and Control by Phillip Hadlington and Christine Marsden (New South Wales University Press, 1998).

Resources

Tiny Game Hunting: Environmentally Healthy Ways to Trap and Kill the Pests in Your House and Garden, New Edition, by Hilary Dole Klein, Adrian M. Wenner, and Courtland Johnson (illustrator) (University of California Press, 2001).

2001 Directory of Least-Toxic Pest Control Products by BIRC (Bio-Integral Resource Center, 2001).

Urban Entomology: Insect and Mite Pests in the Human Environment by William H. Robinson (Stanley Thornes Publishing, 1996).

What's Buggin' You?: Michael Bohdan's Guide to Home Pest Control by Michael Bohdan (Santa Monica Press, 1998).

Wild Neighbors: The Humane Approach to Living with Wildlife by John Hadidian (editor), Guy R. Hodge (editor) and John W. Grandy (editor) (Fulcrum Publishing, 1997).

Wildlife Pest Control Around Gardens and Homes by Terrell P. Salmon (University of California Press, 1984).

PRODUCTS

There are literally thousands of manufacturers of products that will help you to deal with specific or general critter problems in your house or yard. You'll find listings of these companies in many of the books and magazines noted above, or on the Web sites listed earlier. Some of the products you might investigate are:

- Ammonium soaps
- Automatic exploders
- Barriers
- Beneficial insects
- Bone tar oil
- Boric acid
- Botanical pesticides
- Capsaicin (hot) oil
- Electronic alarms and recorded distress sounds
- Fireworks, shell crackers, and whistle bombs
- Flashing or revolving lights
- Frightening devices

TOTAL CRITTER CONTROL

- Glue boards
- Kites, balloons, and raptor effigies
- Live traps
- Netting
- Porcupine Wire
- Putrescent whole egg solids
- Scarecrows
- Sound producing devices
- Sticky substances
- Traps
- Vibrating devices

INDEX

Index

Index

Index

Index

Index

Index

Index

Index

Index